I'm Ready to Talk Three

I'm Ready to Talk Three

VIETNAM VETS PRESERVE THEIR STORIES

AS TOLD TO ROBERT O. BABCOCK

DEEDS PUBLISHING

Published by Deeds Publishing in Athens, GA
www.deedspublishing.com

Printed in The United States of America

Cover design by Mark Babcock.

ISBN 978-1-961505-54-4

Books are available in quantity for promotional or premium use.
For information, email info@deedspublishing.com.

First Edition, 2026

10 9 8 7 6 5 4 3 2 1

This book is dedicated to all those who honorably answered our country's call to duty during the Vietnam War, and all other wars our nation has fought. It is also dedicated to the Family and friends who lived the experience with us and who want to learn more from the point of view of those of us who participated in Vietnam and other wars—past, present, future.

And I especially want to dedicate this book to my fellow Atlanta Vietnam Veterans Business Association members who cared enough about their Family, their friends, their unit's history, and American history to take the time to write their stories to include in this book. It is a legacy we leave for future generations.

FOREWORD

What you are about to read here are a series of stories (remembrances) from a number of men (and women) who either served in the Vietnam War or who waited at home for a loved one who served in that war.

These stories are 'from the hearts' of those who went to what became a very unpopular war because of their loyalty to their Nation. When they returned, they found out that many of their fellow citizens (and the politicians who sent them there) did not appreciate the sacrifices they had made or the friends they had lost.

These men and women survived their involvement with the Vietnam War and went on to live productive lives. They serve as examples of what good citizens should be. They gave their time and effort to their Nation as members of the Armed Forces (or as Armed Forces family members) and then they went on with their lives to become productive members of their communities.

For many, the Vietnam War was a seminal event in their lives and the stories you are about to read will give you some insight as to why their experience in the Vietnam War was the major influence that it was.

President Ronald Reagan said that we were the best America had because we served our Nation when so many others chose not to and to denigrate those who did.

The stories contained in this book will give the reader some truthful insights into the individuals who served in that war and/or to their family members. Enjoy and learn.

Carl "Skip" Bell, US Army Colonel (Retired)
Chairman, Atlanta Vietnam Veterans Business Association

ACKNOWLEDGMENTS

First, I thank the Vietnam veterans and wives who have written their stories to include in this book. As the third book in our I'm Ready to Talk series, you who contributed to our first and second books once again came forward to preserve more of your stories. Plus, several who did not contribute to our first two books realized you had missed an opportunity and stepped forward to be included. I remain impressed with the quality of the way you have expressed your honest memories of what to most of us is the most memorable time of our lives.

I thank my team at Deeds Publishing for making sure this book is as good as it can be. Jan quickly sold me on the title and cover picture (again), Mark used his design talents to make it into a cover that will catch anyone's eye, and Mark and Matt for their layout work.

My editors — Skip Bell, John Butler, Rich Moushegian, and Norman Zoller, teamed their editor's eyes with the experiences they had during the Vietnam war to make sure this book is as real and accurate as it can be.

And I thank all who encouraged me to make this project happen. This is a legacy that all who participated are leaving for generations of Americans to come. I've never had a person tell me they knew too much about the military service of their Family member. This is our gift to our Family, our military unit, and to American history. No political agendas are included in these stories, just our personal experiences.

AUTHOR'S NOTE

With so many authors contributing stories to this book, we (the editing team) have agreed to keep the style and voice of each author in his/her story. Thus, you will see inconsistencies in use of capitalization, abbreviations, military jargon, and style. We considered putting a glossary of military terms in the book, but it is already longer than we thought it would be. We were adamant that we preserved each author's story in their unique style and manner.

Therefore, don't be an English teacher critic if you find some things that don't suit you. Our focus is the story, not making it so vanilla and proper that the story's intent is lost.

If you want to understand a term we use, do like most people do, go to your computer and Google it — or, better yet, ask a veteran. We all speak the same language and would love to explain it to you and maybe tell you another one of our stories because you showed an interest in us. Remember, the title, after years of silence, "I am Ready to Talk." Thus, that is why we titled the book, "I'm Ready to Talk (Book 3)."

Show some interest, ask your favorite veteran a few questions, and I bet he or she will tell you more than you expected to hear.

Also, you will see some profanity sprinkled through the stories. That has been a way of life for all wars since the beginning of time. It's hard to express yourself with, "Gee whiz, that was bad…" Instead, it takes some profanity to best express yourself in these toughest and most memorable times of our lives. If you are offended, it is not our intent to offend but to tell our stories in the best way we know how.

ABOUT THE AUTHOR

BY JAN BABCOCK, REPRINTED FROM I'M READY TO TALK TWO

I must warn you; this story is quite different from the others in this book. It is a peek into the past of a high school girl that knew nothing about the Vietnam War. It didn't make the cut! After all, I had important things to do... cheerleading practice, writing notes to my friends, staring into the eyes of my boyfriend, etc. You get the picture.

Now fast forward to July 1981; the day I married the love of my life, Bob Babcock. This was a second marriage for both of us. We were blissfully happy but extremely poor. Life was good for us as we merged two families together. We had a child together and got into our separate roles as wife, mother, and IBM person for me, and head of household for Bob.

As an IBM executive, Bob travelled a great deal. I still recall his coming home from a meeting in the winter of 1983 and telling me about this guy that he met and got into a very deep conversation about time spent in Vietnam, both with the 4th Infantry Division, but at different times. I should have known to nip that in the bud — just kidding!

In the summer of 1984, Bob came home telling me his new-found friend, Hal Reynolds, had invited him to go to Washington, DC on Veterans Day to attend the dedication of some kind of monument at the "Wall" that had been dedicated in 1982. I knew little about it, but Bob asked me to go with him. I decided, it can't hurt anything. DC should be interesting and maybe I can learn more about Bob's Vietnam experiences, so I agreed to go.

Bob and I together attended our first Veterans' event, the dedication of the 3-Man Statue near the Vietnam Memorial Wall in Washington, DC on Veterans Day 1984. I just remember thinking, what kind of group is this? Men telling stories, crying, laughing, and drinking lots of beer. I listened to them talk and was shaken by some of the things that I heard. You did what???

At one point there was a man, a Vietnam Marine (my father was a WWII Marine — Semper Fi!) that was truly struggling with his past. Somehow, I got in the middle of his anguish, and he grabbed me and wouldn't let go. I was terrified by what I saw in his eyes. It took several veterans to remove me from his grasp. They did it gently, understanding from personal experience what his mind was flashing back to. It scared me; I didn't belong there; this wasn't for me.

But being the supportive, loving wife, I encouraged Bob to continue with his newfound "hobby."

From that first experience, Bob started talking about Army reunions and how he wanted to go. "Fine," I said, "you go, and I'll stay with the children." At that time, little did I know that this would become a way of life for us. (His first National 4th Infantry Division Association reunion was in 1991 when the 4IDA was made up of 90% WWII veterans. He has only missed three since then. Vietnam veterans were just beginning to join and now are the majority of members).

But I understood; I certainly had things that I loved to do, so it seemed natural that he would as well. His involvement in the reunions grew as he became a part of the leadership and executive boards. Bob is a big-picture kind of guy and had huge plans for expanding his Associations' reach to other misplaced Veterans. He kept saying that he wanted me to attend these events. Naturally I had a great excuse; who would take care of the children??

In life, if we are blessed with the ability to follow our passion, we are indeed fortunate. That is exactly what Bob has done; not just for the past 30 plus years, but for the majority of his life. As a boy, Bob constantly played Army. In college he took ROTC and received his commission. That was the springboard for the most memorable year in his life — serving as a rifle platoon leader leading men in combat in the central highlands of Vietnam with his beloved 22nd Infantry Regiment of the 4th Infantry Division. He always said he made more important decisions as a 23-year-old lieutenant in Vietnam than he ever made as an executive at IBM.

From there, life took over, and for Bob, it resulted in a 34-year career with IBM. Remember when I told you about Bob's conversation with a fellow IBMer who also had been in Vietnam? That was 1983 when his passion was ignited… yet carefully tucked away.

As time passed, Bob began attending more reunions, he loved talking with WWII vets and hearing their stories. As he looked forward to hearing more each year, he quickly realized that those WWII Vets were dying, along with their stories.

In an effort to preserve this precious history, Bob took it upon himself to put together a book of 450 stories which included 325 WWII stories, 25 Cold War, and 100 Vietnam stories. Although most of these WWII and Cold War men are now deceased; their stories live on.

Although I didn't attend many of these reunions with Bob, I understood that this was his love and passion. I had no problem with it as I was busy taking care of our children.

In 2003, when the war in Iraq began and Bob had retired from IBM, Bob got heavily involved in communicating via a daily email to family members of the 4th Infantry Division, letting them know what was happening. It was all consuming; sometimes taking 12-18 hours each day and reached 35,000+ people. Now Bob's passion was starting to impact me. At first, I was resentful of the time spent away from our children and me. It was taking a huge emotional toll on Bob; there was nothing left for us.

I'll never forget the day in April 2004 that we were in Ft. Hood at the welcome home for the 4th Infantry Division after their first year in Iraq. Bob was recognized by Maj. General Ray Odierno, the Commanding General of the 4ID (later Chief of Staff of the Army), in front of the large assembly of Soldiers, Family, and friends, with a Commander's Award for Public Service. Bob has said he cherishes that award above any he has ever earned, other than his Combat Infantryman's Badge earned in Vietnam.

After the ceremony, as we were walking to the area where he would be meeting people, we saw a huge line of people, and I wondered what it was for. As we got closer, Bob was engulfed by the crowd. They were there to see him; they wanted to meet this man that was their lifeline to their loved one in Iraq. It was the turning point for me. I finally understand that this was more than a passion; it was what God had whispered in his ear.

I stood back and watched as mothers, fathers, and wives hugged my husband in gratitude for what he had done. I finally understood the importance of his passion. He continued doing that mission for the division's next three Iraq deployments, through 2011.

Because of Bob's extensive knowledge, he is constantly asked to speak to different groups. I never tire of hearing him talk. There is hardly a week that goes by that someone doesn't ask him to help them find out information about a loved one's service record—from WWII to today. He has helped me see the importance of love for our country and the willingness to lay down one's life to keep all of us safe.

So, I want to thank you, Bob, for your service, and for teaching our children and me the importance of standing up for what you believe in. I am very proud of you and for what you have accomplished. You exemplify Deeds Not Words! I love you dearly.

Now read on, you will read stories from many more veterans who are as passionate as Bob is about their patriotism and pride from answering their country's call to serve during the Vietnam War.

VIETNAM WAR: THE GOOD, THE BAD, THE UGLY

BOB BABCOCK, PLATOON LEADER/XO, B/1-22 INFANTRY, 4TH INFANTRY DIVISION – JULY 1966 TO JULY 1967

Vietnam was not a sixteen-year war, it was a one-year war fought sixteen times by a different set of troops each year from 1959 to 1975. Unlike WWII, we were not in for the "duration." The war morphed continually. Each year was different and had its own list of good, bad, and ugly, just as have all wars since the beginning of time.

Our highest troop concentration was in the years 1965 to 1971, with the peak hitting in 1968 with almost 550,000 troops deployed. Earlier and later, we had far fewer US troops engaged. But as my favorite quote from wartime explains, which came from a WWII vet, "Any battle is a big one... if you are in it." So, let's never think the entire sixteen years of the war were any more or less life changing for those who were there.

No person can say, "This is how it was in Vietnam." He or she can only talk about his/her own experiences during the timeframe they were there. Some of the factors that make it impossible for anyone to describe anything more than individual experiences was branch of service (Army, Marines, Air Force, Navy, Coast Guard); type of job (combat arms, aviation, support); rank (private to general); area of country served in (I, II, III, IV Corps, Yankee Station, Thailand, or Philippines); time period served (advisor phase –1959-1964, buildup phase –1965-1967, Tet and major battles –1968-1969, withdrawal phase –1970-1972, Vietnamization advisory phase –1973-1975); public opinion (Ignorance and apathy to mild support, disillusionment, opposition, protests and riots, relief and indifference).

I speak from the viewpoint of being an Infantry lieutenant in the Army in the buildup phase (July 1966 to July 1967), in the central highlands (II Corps), when the American public was apathetic to mildly supportive. I also had the advantage of training my troops while at Fort Lewis, Washington and taking men that I knew to fight next to me.

Just as our fathers from World War II did, we came back from war and lived productive lives. We were mostly ignored until recent years, but random formal and informal polls of Vietnam veterans show that about 90% of us are proud we served in Vietnam and would do it again if called on. There must have been something "good" that causes such a high percentage to feel that way. Among the intangibles that is common among all who have been to war is the brotherhood with those who shared your experiences, whether in the same unit or simply because you fought in the same war.

I will always believe that there is no stronger bond in the world, except for a mother and her child, than that of combat veterans who have fought together. Our fellow veterans have experienced the good, the bad, and the ugly with us — and we focus on the total picture.

* * *

For anyone reading this who do not know a Vietnam veteran to talk to, let me suggest you go to http://www.witnesstowar.org/ and watch video interviews of Vietnam veterans talking about their experiences. From these short (two to six minutes each) video clips, you can get a good view of what we Vietnam veterans think about our experiences in Vietnam. You will find nine separate video clips that Joe Galloway did with me included there — and hundreds of clips of veterans from all branches and specialties and timeframes from the Vietnam War. The stories you will see give a different perspective — they cover the "good" as well as the "bad" and "ugly."

In this book, you will find stories from all years of the war, all branches, all ranks, and we tell it like it was for us…The Good, The Bad, The Ugly.

CONTENTS

TET AND MAJOR BATTLES 71

WITHDRAWAL PHASE: 1970-1972 167

ALPHABETICAL LIST OF AUTHORS

ADVISORY PHASE: 1959 TO 1964

JULY 8, 1959: First two Americans killed in Vietnam.

NOVEMBER 30, 1961: President Kennedy expanded American involvement in Vietnam.

JANUARY 12, 1962: Operation Ranch Hand Began (spraying of Agent Orange in Vietnam).

FEBRUARY 20, 1962: John Glenn orbited the Earth.

OCTOBER 14, 1962: Cuban missile crisis began.

JANUARY 2, 1963: Battle of Ap Bac (VC guerillas defeated a much larger South Vietnam force, American advisors were killed).

JANUARY 11, 1963: Buddhist monk burned himself to death to protest President Diem.

AUGUST 28, 1963: Martin Luther King, Jr gave his "I Have a Dream" speech at Lincoln Monument in Washington, DC.

NOVEMBER 2, 1963: President Diem, president of South Vietnam was assassinated.

NOVEMBER 22, 1963: President Kennedy was assassinated.

MARCH 24, 1964: COL Floyd Thompson captured by VC, longest held POW in American History.

JULY 2, 1964: Civil Rights Act was signed.

AUGUST 2, 1964: Gulf of Tonkin incident: the Gulf of Tonkin Resolution by Congress gave the President unprecedented power to commit US Forces without a declaration of war.

AUGUST 5, 1964: LT Everett Alvarez, US Navy, shot down: became first pilot detained in Vietnam, was released in 1973.

Source: www.vvmf.org

Unfortunately, we did not receive any stories from veterans of this first phase of the Vietnam War. Keep in mind, in these early years we didn't send 19-year-old draftees to Vietnam, we sent our more experienced, higher ranking, and older veterans. Many of them fought in World War II and/or the Korean War. Thus, if they were still alive now, they would be in roughly the 100-year-old time of their lives. To find stories from this era of the Vietnam War, go back to your first book of this series, it has a few stories from this era.

BUILDUP PHASE: 1965-1967

MARCH 2, 1965: Rolling Thunder bombing campaign of North Vietnam began.

MARCH 8, 1965: First brigade sized US force of Marines landed in Da Nang, first large-scale ground force deployed to Vietnam. (Author's note: Jerry West, a guy I grew up with in small town Oklahoma, landed as a Marine on this day).

AUGUST 14, 1965: Advance party of 1st Cavalry Division leaves US for Vietnam.

AUGUST 18, 1965: Operation Starlite began, first major US battle of the war, 5,000 Marines against a stronghold of VC. Two Marines earned the Medal of Honor.

NOVEMBER 14, 1965: Battle of Ia Drang Valley started, first major battle between US and NVA forces.

NOVEMBER 27, 1965: March on Washington for Peace in Vietnam, 25,000 protestors.

MARCH 8, 1966: Australia sent first troops to Vietnam, one of six allies to send troops to fight alongside US. South Korea, New Zealand, Thailand, Philippines, and Taiwan (Republic of China).

AUGUST 6, 1966: President Johnson increased troops deployed to 292,000, which would later peak at more than a half million deployed (Author's note: I landed in Vietnam on this date: will never forget it, as will all our vets remember when they arrived).

OCTOBER 21, 1967: Over 50,000 protestors marched on the Pentagon in protest of the war.

NOVEMBER 30, 1967: American deaths in Vietnam hit the 15,000 mark.

Source: www.vvmf.org/VietnamWar/Timeline

JOHN E DOUGLAS, SR.

BRANCH OF MILITARY AND JOB HELD: Army Transportation
Officer — Company Operations Officer
DATES OF SERVICE: 1964 — 1970
UNIT SERVED WITH IN VIETNAM: 151 Light Truck Company, 27th TC
Battalion Truck
DATES SERVED IN VIETNAM: Sept 1, 1965 — June 30, 1966
HIGHEST RANK HELD: Captain (O-3)
PLACE AND YEAR OF BIRTH: Atlanta, GA — 1942

THE TRIP OVER, ETC

August 14, 1965, I kissed my wife and boarded a chartered Pan Am flight from Ft. Campbell, KY to California. After landing, we were taken by bus to Oakland Naval Base, Oakland California. There we boarded the USNS General W. H. Gordon (T-AP-117). The Gordon was a troop transport that served in WWII and later the Korean and Vietnam Wars. It was built in 1944 as a transport for the Navy. It was decommissioned in 1970. Needless to say... it was NOT a luxury liner.

We arrived midafternoon and were greeted by Red Cross "Donut Dollies" and given donuts as we boarded the ship. I was one of 20 2nd LT's (lieutenants) assigned to a cabin that had three bunks high and one porthole.

The Lt's were from all types of Army units. Another LT and I were from the 151 TC (Lt. Trk. Company). The others were from various Signal and Quartermaster units.

Our enlisted men were assigned to a specific area five or six bunks high in the "Hole" of the ship. Officers were told we could have shore leave but must be back on board by midnight that day. We were sailing at 6 a.m. on August 15, 1965.

At midnight, 19 of the 20 LT's were in our cabin. One Signal Corps LT was missing. I will just call him "Red". As we stood on deck the next morning watching the shore personnel untying the ropes and preparing to remove the gangplank, a small green sports car screeched to a stop and "Red," along with a young strikingly beautiful blonde jumped out of the car. "Red" gave her a large bend over backward kiss and ran up the gangplank just as it was being raised. Of course, he bragged for days that he was the last of the 5,000 men on board to have "been" with a woman. More about "Red" later.

As we pulled away from the dock and cruised through San Francisco Bay, standing on the bow of the Gordon passing Alcatraz Island and going under the fabulous Golden Gate Bridge, I thought to myself "this will be a great 17-day cruise."

Shortly, as we left the inlet and encountered the "breakers," the ship began to rock and sway. I went straight to my bunk and took a Dramamine. I stayed in my bunk for two days until the sea calmed. I also learned when I went to the dining room NOT to look out the portholes as I was eating. If you did, you were looking at the water one second and at the sky the next minute. Very easy to again feel nausea.

We were instructed to take a Navy shower. Only shower once every two days. The instructions were 1. Get in the shower. 2. Turn on the water and get wet. 3. Turn off the water. 4. Soap up. 5. Turn water on and rinse. 6. Get out of the shower and dry off. This whole process was to be accomplished in two minutes or less.

Another story about "Red." Liquor was not allowed on the ship. Of course, "Red" had a whole suitcase full of liquor and could be seen taking a sip from time to time.

Before I had left Fort Campbell, my mother-in-law asked if

she could print my letters in the three weekly northeast Georgia newspapers she owned and published. My wife had an aunt that was a Presbyterian missionary to Brazil and for months they published her letters under the title "Letters from Martha." She envisioned a similar column from me. I told her it was Ok with me, but she would need to "clean them up" because I would not be writing them to be published. She assured me she would "Clean them up."

Since there was no mailbox to mail letters, I wrote a continuous letter for 17 days to my wife and mailed it the first chance I had once we reached Qui Nhon. Thirty days later, I began to receive mail. One of the first pieces of mail was a copy of The Lavonia Times, one of the papers my mother-in-law published. I was enjoying reading about the local news when I opened the centerfold and there was my letter- two full pages, word for word, to include my description of the horrible conditions my men were living in in the Hole. Complete with the stench of vomit and urine. It was a vivid and descriptive narrative that went on for several paragraphs. Luckily for me, I guess no one in the Pentagon read The Lavonia Times that week.

When we arrived in Qui Nhon, which is a shallow water port, we had to disembark by crawling over the side of the ship and climbing down the rope netting "ala D-Day" into the waiting landing craft, not knowing what to expect when we landed because this was very early in the war.

The temperature was 120 degrees that day. A couple of months later, our Battalion (27th BN TRK) erected a sign on that same beach that read **"WELCOME TO VIETNAM! This beach taken and secured for your comfort and protection by the 27th Transportation Battalion."**

We stayed in pup tents for the first week while we erected squad tents, and eventually erected a Mess tent.

One scary time was our first night. We had not received our

weapons. The only weapon we had was my personal .22 caliber short pistol that I had hidden in the company safe, since I was the Secured Document Officer.

We ate out of cans for seven months, except for Thanksgiving and Christmas, until we finally scrounged a couple of freezers.

One last "Red" story. I ran into one of the Signal officers that was in the same company as "Red". He was a unit commander of one of the relay signal units that were stationed on top of the mountains and relayed messages up and down the coast… one of the units that received mail and food once a day by helicopter. One day their Battalion Commander decided to pay them a visit on the daily chopper run. When the Colonel arrived, a SFC greeted him and saluted. The Colonel inquired where was Lt. "Red"?

The SFC responded that he did not know where he was. About three weeks before, "Red" boarded the daily chopper when it left the top of the mountain and he had not been seen or heard from since. An All-points bulletin was sent out and a few days later "Red" was found, "shacked up" with a woman in Saigon. Of course, he was court-martialed and sent back to Ft. Leavenworth, KS (the infamous Army prison).

MY 1ST CONVOY, SEPTEMBER 1965

I became friends with LT Burnett the first week I was in country at Qui Nhon, Vietnam. He was a 2nd LT, the same as me, but he had been in country about two weeks earlier. I arrived on September 1, 1965. He was with the 58th TC Lt. Trk. Co. and I was with the 151 TC Lt. Trk. Co. While our trucks (2 ½ Ton) were being unloaded, I volunteered to ride with LT Burnett to learn the ropes and understand our mission and the route we were to use. We were beginning to move the 1st Air Calvary into An Khe. That was about 50 miles from Qui Nhon, via Hwy 1 and Hwy 19.

There were four of us in the jeep. LT Burnett and his driver were up front with a SGT manning the 30 Caliber Machine Gun mounted "Rat Patrol" style in the rear. I was sitting on top of the right rear wheel cover. The jeep was covered in sandbags to try to minimize damage from land mines/IED.

There were seven bridges on the route that had been blown up, but the Corps of Engineers replaced them with portable Bailey Bridges. The 101st Airborne was responsible for security on those bridges and Hwy 19 until the 1st Air Cavalry had time enough to establish their base camp and receive their equipment.

I felt very secure with the 101st Troopers. Later, when I began running the road, they would snap to attention and ask how many vehicles I had. They would count them and radio me the count at every bridge.

Of course, I was anxious, being my first day on convoy in country. To make matters worse, sand blew in my right eye and I could not get it out, even after pouring half the water in my canteen in my eye.

Finally, I got it out when we were near the base of the An Khe Pass. That day and for several weeks, when I would arrive in An Khe, we would literally be following the bulldozers as they cleared the road around what would eventually become known as Camp Radcliff or, locally, as The Golf Course.

We carried concertina wire, C-Rations, and ammunition that day and for weeks later before also extending our route eventually through the Mang Yang pass into Pleiku.

That first day was relatively quiet, unlike later convoys where we experienced sniper and small ambushes. We had to learn that everybody in black pajamas was not the Viet Cong.

Upon return, I briefed our drivers about what to expect and how to prepare for the convoys with weapons, food, and what to watch for, as well as how to maintain convoy discipline. Don't get too close to the other trucks.

LT Burnett and I remained friends. My wife and his wife meet for lunch near the GA/SC line several times a year.

ROBERT O. "BOB" BABCOCK

BRANCH OF MILITARY AND JOB HELD: Army, Infantry Rifle Platoon Leader and XO
DATES OF SERVICE: 1965 to 1975
UNIT(S) SERVED WITH IN VIETNAM: Bravo Company, 1-22 Infantry Regiment, 4th Infantry Division
DATES SERVED IN VIETNAM: July 1966 to July 1967
HIGHEST RANK HELD: Captain (O-3)
PLACE AND YEAR OF BIRTH: Heavener, OK - 1943

DON'T USE THE LAUNDROMATS / THE GREAT NORTHWEST

A green second lieutenant, newly graduated from college with an ROTC commission, I spent June and July 1965 doing what I had done during my college summers—working for my dad in his trophy making shop in small-town Oklahoma. But my mind was elsewhere, ready to start my two-year commitment as an Infantry lieutenant. Fort Benning, Georgia was my starting point.

As my wife, Phyllis, and I packed our new maroon 1965 Mustang to head south in late July, President Lyndon Johnston announced, "I have today ordered to Viet-Nam the Air Mobile Division and certain other forces which will raise our fighting strength from 75,000 to 125,000 men almost immediately." This marked a significant escalation, with the deployment of the entire Air Mobile Division.

It was interesting, but not significant to me (or so I thought) as I was headed after Infantry School to Fort Lewis, Washington to join the 4th Infantry Division. Our mission was to guard the great Northwest portion of the United States.

The first thing we saw when we pulled into Fort Benning were railroad flatcars being loaded with the vehicles and heavy equipment of the 1st Cavalry Division. The next morning, the front-page headline in the Columbus, Georgia newspaper was, "DON'T use the laundromats!" The story that followed explained that the entire 1st Cavalry Division was dying their white underwear with green dye and that would take a long time to go away.

In early September, I came home one evening and Phyllis said, "You've got to watch tonight's news! Renny Reeves was interviewed in Vietnam!" I never stayed up past 9:00, needed my rest for the tough training each day. But that night I stayed up until 11:00 and watched Renny, a friend from college and a year ahead of me in ROTC, talk about coming out of helicopters with the NVA shooting at him. I clung to my belief that my mission would be guarding the great Northwest.

In mid-November, after finishing Infantry and AIRBORNE Schools, we were driving to Fort Lewis and listening on the radio to the regular updates about the Battle of the Ia Drang Valley that the 1st Cavalry Division was fighting that week with the North Vietnamese Army (NVA). Although I still thought guarding the great Northwest would be my job, I wasn't nearly as confident as I had been a few months earlier.

By mid-December, we were preparing to "train and retain" hundreds of fresh out of Basic Training new Soldiers. Although Vietnam was not in our orders, we quickly understood that we had a very serious job to get ready for — and it had nothing to do with the great Northwest. (See my *Training at Fort Lewis* story in *I'm Ready to Talk Two*.)

JOHN E DOUGLAS, SR.

A LUCKY SOLDIER

The below are excerpts from an article written by The Atlanta Constitution, writers Remer Tyson and Harold Martin about Col (Ret.) Jack Peevy. Jack Peevy was a Lifetime Member of the AVVBA, passing away on Jan. 7, 2014.

Jack was a 1964 graduate of North Georgia College, now known as the University of North Georgia. He and I served in Vietnam at the same time, but in different units. In 1965, he was a Rifle Platoon Leader in the 1st BN (Airborne), 8th Calvary and 1st Air Calvary Division in Vietnam. He was wounded on December 18th, 1965. The article said he received six wounds that day, but he always told me it was seven wounds. He used to say I was wounded seven times, but I only received one Purple Heart.

I came in off convoy to Pleiku that day and was met by our 1st Sgt. telling me there was a Lt. Peevy that was in serious condition at the 85th Evac. Hospital in downtown Qui Nhon, and he was asking for me. By the time I could get to the hospital, he had already been air evacuated to the Philippines. So, we did not reconnect for several years after I returned to the USA.

THESE ARE THE QUOTES FROM THE NEWSPAPER ARTICLE:

First Lieut. Lewis J Peevy of Alpharetta GA, recuperating from six wounds sprayed from his chest to his ankle, said he felt "guilty" about spending only three months in Vietnam and wants to go back.

Peevy, 24, hobbled around on crutches in the hospital corridor awaiting a Thursday afternoon flight home to Georgia. He hopes to spend six weeks with his folks in Alpharetta and hopes he will be physically sound in a few months for another combat tour in Vietnam. "I didn't spend but three months over there," Peevy said, "I feel kind of guilty about it."

Peevy, as a Platoon Leader on a brigade-sized search and destroy operation near Bien Khe last Dec 18, was wounded when bullets slammed into his chest, arms, lower left leg, left hip, and right thigh. Peevy's feeling of "guilt" goes back to the fact he got off light that day with his six wounds. Two of the men in his platoon were killed and six were wounded.

Since he has been in the hospital, the Lieutenant has seen how lucky he was. Another member of his platoon, wounded the same day, is paralyzed from the waist down.

EXCERPTS FROM ANOTHER ARTICLE WRITTEN BY HAROLD MARTIN QUOTES HIS MOTHER:

Her voice was very low and calm and matter of fact. Her son, she said, was in Walter Reed Hospital in Washington. D.C., and she would like to send him the Atlanta Constitution. He would probably be there at least three months, maybe longer (actually stayed about one year). He had been severely wounded in Vietnam. "The thing about it is—I think I saw the battle in which he was wounded," she said, "It was on TV. It named the outfits that were involved, and one of them was his. I did not see him on the screen, but I did see that there were men who had been wounded and there were some men who had been killed."

So, the next few days were very bad days for me. I dreaded

answering the door and finding a messenger standing there with a telegram in his hand. But the days passed, and the telegram didn't come, and I began to breathe a little easier.

Then two days after Christmas, I got a letter from him. He had been in that battle after all and had been badly wounded. He had been hit in the face, and in the chest with shrapnel, and he had been hit just at boot top level by automatic weapons fire that had almost torn off his left foot.

He asked the Defense Department not to notify me by telegram for he knew it would spoil my Christmas, just to have the bare notice that he had been wounded. Instead, he wrote me himself from the evacuation hospital explaining everything. And he had the letter mailed so it would get to me after Christmas.

He is a First Lieutenant. The Army is his life. All he wants now is to be healed of his wounds so that he can go back to lead his platoon again. I told her, as we finished talking, how much I admired her courage, her matter-of-fact acceptance of the career her son had chosen.

"I think I learned from him," she said, when I wrote him telling how sorry I was about his leg," he said, "**Look, mother, you don't understand. I'm the lucky one. I've still got all my arms and legs. And lots of the guys in this ward haven't. I am the Lucky One.**"

FURTHER INFORMATION ON LT. PEEVY

Jack Peevy retired as a Colonel after 30 years' service. His awards and qualifications include the Legion of Merit, Bronze Star with V device, Purple Heart, Meritorious Service Medal, Parachutist Badge, Combat Infantryman's Badge, and the Special Forces Tab.

After retiring from the US Army, he became a successful real estate entrepreneur in North Fulton County. He left approximately $5 million dollars to the University of North Georgia in support of the Boar's Head Brigade. In addition, he established the Peevy Scholarship for nursing students at North Georgia, because nurses helped him so much over the years.

While he was recovering in a ward at Walter Reed Hospital, he met Army helicopter pilot Steve Martin and later they reunited at the Atlanta Vietnam Veterans Business Association.

ROBERT O. "BOB" BABCOCK

"TRAIN AND RETAIN"

As signups were announced for AIRBORNE and Ranger schools in the last weeks of Infantry Officer Basic Course (IOBC) at Fort Benning in September 1965, our drill instructors had a surprising announcement. They said, "You guys wearing the 4ID patch can sign up for AIRBORNE school, but they need you out at Fort Lewis by December… so you can't attend Ranger school now. Maybe later."

I was thinking about Ranger school but was sticking to my decision made in ROTC before I graduated. I had turned down my Regular Army (RA) commission that I had earned from being ranked high in my class. My logic was, if I like the Army, I can apply for RA later; if I don't like the Army, I can get out after two years instead of three. This turned out to be one of the smartest and luckiest decisions I ever made.

Unlike my early thinking, the Army didn't need the 4ID to guard the great Northwest, we were needed in a place called Vietnam. As the war went on from 1965 to 1975, officers with a Regular Army commission were frozen and served longer than the three years they had agreed to. Many, if not most of them, had multiple tours in Vietnam. After my one-year tour in Vietnam, I ETS'd (Estimated Time of Separation) when I got back to the States in July 1967.

The great advantage we had with "train and retain" in 1965-1966 is we brought in new Soldiers, fresh out of basic training, and gave them "in unit" Advanced Individual Training (AIT). As their leaders, we learned their capabilities by training them. Who were the best marksmen with various weapons, who were good at reading a map and following a compass' direction, and even more importantly, who were the leaders.

We went to Vietnam in July 1966 with every fire team leader slot filled by a draftee we had trained, and over half of our squad leader jobs were filled by drafted troops who had bubbled to the top during our six months of training together at Fort Lewis.

Contrast that to later in the war when Soldiers flew out to their unit on a helicopter, spent their first night in a foxhole with someone they had never seen before, and the leaders had no idea who the best people were among the replacements.

The Army learned their lesson in Vietnam. Today's Army trains together in the U.S., deploys together, and comes home together. Train and retain—a very valuable "lesson learned" that came from the Vietnam War.

ELIZABETH (BETTY) LITTLE DOUGLAS

MARRIED TO JOHN E DOUGLAS SR. ON JUNE 6, 1964
HAPPILY MARRIED FOR 61 YEARS

CHRISTMAS SURPRISES

Our second Christmas as a married couple was going to be spent halfway around the world from each other. I am sure that John and all the soldiers were aware and wondering what this holiday would involve, but I was living with John's family while he was gone, and I know that all of us were quite aware of the approaching holiday. Our physical bodies were in Atlanta, GA, but our hearts and thoughts were far away in Vietnam.

We had gone and purchased a small Christmas tree to send to John to try to brighten up a tent and had spent many nights popping corn and preparing decorations for that little tree. The pictures he sent back of that little tree looked like a "Charlie Brown" tree, but I promise you it was filled with love and longing from all of us.

This unusual Christmas was approaching; we had decorated the house and bought and wrapped presents for each other, and my mother and brother had arrived from Lavonia so that we could all be together to celebrate. All of a sudden, John's cousin, who was a florist in Griffin, GA, shows up and enters the house with great cheer and excitement—more than any of us had been able to muster.

She goes out three different times and comes back in with the huge arrangements of red roses—the prettiest, most wonderful roses any of us had ever seen. She had obviously selected these flowers with great care. Each arrangement had a different name, one to me, one to my mother, and one to John's mother. What in the world? We could not understand. It turns out John had sent a money order for $30 and ordered 3 dozen roses from his cousin to be delivered for Christmas.

She more than filled that order, delivering those flowers all the way from Griffin, making sure the arrangements were perfect and making it into such a celebration. I tear up as I think about that day. What a surprise!

I took pictures of those roses and showed them to everyone—when I showed the people at work, one girl said she wished she had a husband that would send her roses! I just stared at her. Did she realize what she was saying? Those roses lasted the longest of any I have ever had and opened all the way out—they spoke volumes every day when I walked into my room after work, and they were the last thing I would see at night.

JOHN E DOUGLAS, SR.

ANN-MARGRET USO TOUR – SPRING 1966

I was contacted by the Battalion Commander and asked if I would coordinate the first USO show to come to Qui Nhon. The show was Edgar Bergen and his puppet Charlie McCarthy. I was referred to a couple of Special Services Officers for the details. After getting the details, date, time, etc., I arranged to borrow plywood from the Engineers, housing from the Corps Commanders Headquarters, MP Security, publicity, etc.

On the day of the event, the Battalion Commander and I went to the airfield to meet the flight. We waited and waited and waited. Finally, the Special Services officer announced the show had been cancelled. In fact, the show had been cancelled a week earlier and the Special Services Officer failed to notify me. The Battalion Commander gave them a good ass chewing.

A few weeks later, I was asked to coordinate the next show, which was Dean Jones. Fortunately, we were notified a few days earlier that show had also been cancelled.

Once again, I was summoned to the Battalion Commander's office. He apologized about both other shows but wanted to give me first refusal on the next show. The Ann-Margret Show with Johnny Rivers as the co-star. Of course, I jumped at the chance. I again coordinated and organized all the details.

The Colonel and I met them at the airfield. She was beautiful and friendly. She had a Special Forces Captain accompanying them all over the country. I was just the Escort Officer for her show in Qui Nhon. I had arranged for Ann-Margret to stay in the II Corps Chaplain's room, the nicest room in Qui Nhon. The General lived inside the Headquarters, so his room was not available. I asked the Captain about Johnny Rivers and his two backup musicians. He said just put them in the BOQ and give them two bottles of whiskey.

The next day, we visited the 85th Evacuation Hospital, visiting with patients and signing pictures of her from a recent movie. Later that day, the show was to be held in the motor pool of the 2nd TC Med. Trk. Co. Before the show, we had arranged for a steak dinner and for her to cut the ribbon for our company's new mess hall. We scrounged the steaks off one of the Navy ships in the harbor.

I had arranged three flatbed trailers, parked side by side, with plywood floors nailed to the trailers and a six-foot bamboo mat across the stage. The Battalion Headquarters furnished the microphone and speakers. When I introduced Johnny Rivers, I heard a lot of boos because they wanted Ann-Margret. They warmed up to him as he sang a couple of his hits.

I went back to check on Ann while Johnny was preforming. She said she needed to use the restroom. I had anticipated that and had an Officer's Latrine prepared, complete with sheets covering the screen wire walls. There was a single light bulb hanging from the ceiling. She asked me to come inside with her while she got her bearings and then for me to turn off the light because she was scared she would cast a shadow on the sheets.

Later she asked for a couple of minutes to be left alone. She was suffering from a sore throat. Every time we stopped, I had hot tea and honey to coat her throat. After the tea and honey, she sat alone, almost as if she put herself into a trance. She then went out and gave a great show.

Forty something years later, I saw that she was coming to Atlanta to star in the *Best Little Whorehouse in Texas* at the Fox Theatre. Several weeks ahead, I reached out to the President of her Fan Club to see if I could meet her and give her some old pictures from her time in Qui Nhon. He asked for me to contact him a week before and let me know which show we would be attending. He arranged for my wife and me and another couple to go backstage after the show to see her.

When she came out in the hall and I introduced myself to her and said I had some pictures for her, she went back and got her glasses. We stood in the hall and "oohing and awing" over the pictures, commenting about her hat and dress, etc. At some point I started to say, "I am sure I am not the first, or will I be the last, to tell you how much it meant for you to come and perform for us."

Right in the middle of my sentence, I broke down and the tears started to flow. She came right over and gave me a great big hug and said, "how glad she was that I made it home."

She is a beautiful and compassionate lady!

EDWARD ETTEL

BRANCH OF SERVICE AND JOB: US Navy, Line Officer
DATES OF MILITARY SERVICE: June 1963 - October 1989 (Combined Active Duty and Reserve)
UNIT SERVED WITH IN VIETNAM: 765th Transportation Battalion; HQ, 1st Aviation Brigade
DATES SERVED IN VIETNAM: December 1971 - November 1972
HIGHEST RANK HELD: Major
PLACE AND DATE OF BIRTH: Corpus Christi, Texas, 1940

BOOZE AND CHOPPED LIVER: OCEAN LOGISTICS SUPPORT OF THE VIETNAM WAR

Supplies for the Vietnam War moved primarily by ocean ships. Between mid-1965 and late 1966, approximately 90% of all cargo was supplied by ship. Early in the Vietnam War, troop ships carried two thirds of U.S. troops to Vietnam; but after the troopships were retired, airlift was used to move the great majority of troops. Priority cargo was also airlifted, but it accounted for only a small fraction of the total tonnage moved.

After serving aboard the USS Hollister (DD-788) in the Pacific and South China Sea in 1963-64, as a LTJG, an O-2, I was ordered to the Military Sea Transportation Service Office Manila which was later renamed Military Sealift Command Manila (MSC Manila) to match the more glamorous Air Force's Military Airlift Command. My only previous Philippine experience was ship liberties in Subic Bay, with its raunchy party-centric town, called Olongapo.

My initial duties were Assistant Operations Officer, Passenger Officer, and Boarding Officer. My office was in the TRANS-CO terminal building on a pier near downtown Manila, so my wife Mary and I moved into an apartment in downtown Manila. Therefore, we lived on the economy, not on a military base, because the closest base, Sangley Point Naval Base, was across Manila Bay and too far away to commute to MSC business in Manila.

Prior to the Vietnam War, Military Sea Command (MSC) operations in the Philippines were relatively calm and easily managed, which was the operational state when I reported for duty on 12 January 1965. The departing officers described the office operations as an easy routine. That pace changed with the introduction of U.S. combat troops to Vietnam beginning on March 8, 1965. Soon the office was handling as much cargo tonnage in a month as had been done in the entire year in 1964. In addition, the office was doing so with one less officer, the Executive Officer (a Lieutenant Commander) having been temporarily transferred to Vietnam to set up an MSC office there.

This was part of a plan by MSC Far East and MSC Pacific to quickly put 50 staff members into Vietnam by taking them from other MSC offices throughout the region. The Vietnam operation could not wait for the eventual increase of MSC's total complement of men.

Because the Manila office was without its Executive Officer, watch officer duties, previously split three ways between the Executive Officer, Operations Officer, and Assistant Operations Officer, were now split just two ways between the Operations Officer and Assistant Operations Officer. The Commanding Officer did not assume Watch Officer duties.

MSC is responsible for ocean shipping of all U.S. Government cargoes throughout the world and was already the largest ocean shipping operator in the world before the escalation. Although not as fast, ninety-five percent of all world-wide military cargo goes by sea, not by air. Military cargoes that were handled included munitions, marine and aviation fuels, vehicles, weapons, base supplies, construction materials and equipment, household effects and personal items. Nonmilitary cargoes included items in support of various U.S. government agencies such as the U.S. Embassy, USAID programs, and the Peace Corps.

Soon Vietnam supply requirements grew such that too many ships were anchored off Vung Tau, Qui Nhon, Da Nang, and Cam Ranh Bay. This became a security, economic, and readiness problem, so a smart decision was made to stage the ships in the Philippines instead of waiting offshore Vietnam. In Philippine anchorages the ships were secure, making hazardous duty pay unnecessary. Also, ships' provisions could be maintained, fuel tanks topped off, and maintenance performed.

In addition, the crews were happier because they could have great liberty. Soon there were over 300 ships positioned in the Philippines, with only a 36-hour cruise over to Vietnam. But we were very busy maintaining control and support of this volume of ships. At this time, my fellow officer, LTJG Mattox, received retirement orders, and so I assumed additional duties as Operations Officer and Cargo Officer.

Early every morning, I would board the Philippine Pilot boat and board newly arrived ships at anchor in Manila Bay, meeting the captain, providing port status and information, giving ships orders and satisfying requests. It was a civilian ship custom for the captain to offer a beverage as he and I conducted business. This was usually some type of alcoholic drink, scotch, bourbon, whisky, or, in the case of Japanese captains, sake.

This protocol could be a problem after visiting over eight ships, especially on hot days climbing ladderways, but someone had to do it. I started taking snacks with me to help absorb the alcohol.

The Philippine Quarantine, Customs, and Immigration officials also boarded the ships, but used their converted tugboat that was much more stable and had a lot more space. They needed more space for their booty, because another 'custom of the port' was for each captain to give each official a bottle or two of booze. At the end of the day, instead of going back to their dock, they would motor up the Pasag River and unload their booty into their private warehouse. Of course I had no such privilege.

Sometimes I would come back to the dock late at night. One time I was walking down an isolated pier, a week ahead of Philippine elections, when automatic weapons opened up and I heard bullets whizzing around the pier, and around the wall of the warehouse. I ran toward the side of the warehouse and jumped into and behind a large pile of trash. Of course I was wearing my tropical white long uniform, and Mary was not happy when I came home very dirty and smelly, but in one piece.

The Philippines could be dangerous around election time, with rival factions resorting to violence. They were not targeting me but were shooting up the place for some reason I never knew.

MSC was also responsible for transport of U.S. government

personnel (military and civilian) who could not fly due to medical or other reasons but could be designated for ocean travel. These included high ranking officers and officials being transferred in and out of the Philippines and Southeast Asia. So, my favorite duty while in MSCO Manila was Passenger Officer, which involved being responsible for passenger operations for Southeast Asia.

I billeted passengers and troops for onward passage to Guam, Okinawa, Yokohama, Honolulu, and San Francisco; and for passenger cruise lines I booked them for onward passage to Hong Kong, Tokyo, Honolulu, and San Francisco. One MSC troop transport called on Manila and Subic Bay once a month, and many military personnel would be given orders to return to the States via MSC troop transports, which included five un air-conditioned ships (the USNS Marshall, USNS Mitchell, etc.) and three air-conditioned ships (the USNS Darby, USNS Barrett and USNS Geiger, which were acquired from American President Lines (APL).

Besides troops, many Filipinos who were retired from the US military services and regular US military, mostly enlisted personnel, were eligible to go space-available on these ships with their families. The air-conditioned ships were greatly favored by the troops, and space-available people as well.

The Barrett was nicknamed the 'Grin and Barrett', which I thought was unfair because it was air-conditioned. I also booked lucky personnel and their families on American President Lines (APL) ships, the SS President Roosevelt, the SS President Wilson and the SS President Cleveland. All were a combination cruiseship and cargo ship, and it was a special prize to be able to be billeted on one, especially for the 23-day voyage back to San Francisco via Hong Kong, Tokyo, and Honolulu, considered as 'Proceed Time' and not vacation time.

I would receive long booking lists from all the South China Sea area ports to book passengers and troops on the ships and then assign them specific cabins or troop spaces based on many factors. I won't describe this process because it was very complex. Instead, I will describe the fun part of the process.

Wearing my Service Dress Whites, I would catch the Philippine Pilot boat much earlier in the morning before sunrise, and when the cruise ship would pass Corregidor Island and enter Manila Bay, it would slow to about eight knots, When the ship was about a thousand yards away, my pilot boat would do an intercept in order to come along the port side of the ship. The ship would go as slow as she could, however my putt-putt pilot boat could not go very fast, so we only had about 30 seconds for the pilot and me to climb about twenty feet up a Jacobs ladder, a flexible hanging ladder consisting of vertical ropes supporting horizontal rope rungs.

Normally Pilot ladders are often incorrectly referred to as Jacobs ladders, with specific regulations on step size, spacing, and the use of spreaders, long treads that extend well past the vertical ropes to stop the ladder from twisting about its long axis (possible when a ship rolls and the ladder is no longer in contact with the ship's side) with the person possibly becoming trapped between the ship's side and the ladder. In our case, I am pretty sure the ladders we used were not as fancy.

While the pilot was climbing up the ladder, the ship's crew would throw me a line and I would tie a knot around my ugly brown satchel's handles, and they would haul the satchel up, which contained all the tickets and other papers. By that time, timing my jump just right with the waves, I would grab the rungs and start climbing, avoiding the Pilot's heels as I did so. This was always a thrill with the ship underway and all, but I found out later that this was also very dangerous, and probably the most dangerous thing I did on active duty.

Timing the jump, with waves, the pilot boat veering back and forth as well as up and down and bouncing off the ship was just part of it. It was also not easy to climb the moving ladder.

After several months of doing this three times a month, I asked why the Quarantine, Customs, and Immigration officials boarded after the ship was docked, and not do as the pilot and I did. It would speed up the whole process.

They told me that a year ago two Customs officials were killed when they fell from the ladder, and because they had to fall right next to the ship, they were sucked into the propellers and ended up resembling chopped liver. After that incident, the officials refused to climb the Jacobs ladder. I then realized why so many people were hanging over the ship, watching us climb the ladder, and I was much more careful from then on, doing this for two more years without incident.

The best part was that once aboard, I was a hero. Not because I avoided the ship's propellers, but because I held the golden tickets and paperwork for families that would be boarding the ship when the ship tied up to the pier. I would then meet each family as they boarded, describe the process, give them their tickets and lead them to their cabin.

They were always very, very excited, and thought I was the cat's meow. It was like they thought I had paid for their 23-day trans-pacific cruise myself. I felt like Santa Claus, and they couldn't stop thanking me, and usually had someone take our group photos.

However, I was not very smart, because this was the time to trade business cards and say, "Have a wonderful cruise. Please keep my card and someday we might cross paths." Because in June of 1967, I was separating from active duty, did not have a job, and could have used a favor or two.

ROBERT L. "BOB" HOPKINS

BRANCH OF MILITARY AND JOB HELD: Army, Infantry Rifle Platoon Leader

DATES OF MILITARY SERVICE: 1965 - 1968

UNIT(S) SERVED WITH IN VIETNAM: 2nd Battalion/39th Infantry, 9th Infantry Division

DATES SERVED IN VIETNAM: December 1966 to December 1967

HIGHEST RANK HELD: First Lieutenant (O-2)

PLACE AND YEAR OF BIRTH: Baltimore, MD — 1943

JUNGLE SCHOOL

The certificate reads that Second Lieutenant Robert L. Hopkins has successfully completed the Jungle Operations Course at the United States Army School of the Americas. It was awarded at Fort Gulick, Canal Zone (Panama) on 2 September 1966.

Before being deployed to Vietnam, the Army wanted to give us a taste of what the environment would be like, so some of the luckier troops were sent to "Jungle School." From Bravo Company, Second Battalion/ 39th Infantry Regiment, Ninth Infantry Division, four of us were selected for jungle training. Although the certificate says we were at Fort Gulick, Canal Zone, I remember being at Fort Sherman in the Canal Zone.

Those selected for the training boarded a DC-3 twin-engine, tail dragging, propeller plane at the Manhattan, Kansas, Municipal Airport to travel to the Air Force Base in Charleston, South Carolina on 21 August 1966. The flight was memorable because the plane was old. Every time the engines accelerated, a misty white cloud would come out of the overhead air conditioning vents, filling the cabin with a dense fog. At Charleston, we boarded a Braniff Airways big yellow banana for our flight to Panama. One of the joys of flying Braniff was the "hot pants" the flight attendants wore.

The plane landed at the Naval Surface Warfare Center military airport on the west coast of Panama, close to Panama City. The group was directed to a bus that took us to an old steam train to cross the isthmus within the Canal Zone. We left the train, boarded a deuce and a half truck. The truck crossed the Panama Canal before arriving at Fort Sherman, our destination. Fort Sherman was located at Toro Point at the Atlantic entrance to the Panama Canal and is close to the city of Colon. The fort was deactivated and turned over to Panama in 1999.

Fort Sherman consists of six, two-story barracks made of white stucco, and they were opened to the elements, having no windows, or air conditioning. The 12-day course we were starting emphasized the skills to fight in a jungle environment and included dismounted mobility, land navigation, jungle tactics, and survival training. When I checked in, I was issued a very rusty old M-14 rifle that was to be mine for the duration of the course.

The next morning, training began with an orientation. Our first assignment was to hike into the jungle and build a bohea, which would be our individual sleeping quarters for the week. The bohea was an above ground platform with overhead cover where one was to sleep at night. The elevation was for two reasons: first it kept you above the damp ground and secondly, it kept the low crawling jungle critters from attacking. The bohea was to be constructed from only natural material. This meant looking around you and creating a sturdy shelter. I can recall the first night in my bohea, everything was quiet until there was a loud crash, then an expletive as someone's bohea collapsed. We spent the next three nights camping out in our boheas.

Since we were already in the jungle, we started to acclimate ourselves to the surroundings. The first exercise was learning to navigate on the land using a compass and pacing our distance. This became an extremely useful exercise, since this was our method of land navigation in Vietnam. Lesson two was making a float for a water crossing. The Chagres River flowed through the Fort Sherman reservation, becoming the ideal obstacle for us to use as a training aid. We had one of the stronger swimmers take a rope across first, then we were to cross with our floats. We made our floats by laying our ponchos out then filling them with brush and tying them up in a bundle. The ponchos made an ideal float to use as an aid in fording the river.

While we were in the field, it became time for rope training. This training was about using ropes to rappel off a cliff. It was late one afternoon when the instruction began on a gentle slopping dirt cliff. Unfortunately, I never got the opportunity to practice since the training day ended and we went back to camp. The next day we were to go to a more demanding cliff to qualify for our "Jungle expert" designation.

As expected, we had a typical tropical rainstorm as I was rappelling for the very first time. Never in my life was I as frightened as I was when rappeling off a rock cliff that had a reverse slope and a waterfall running down it. Until I got the knack of how to rappel properly, I held on for dear life with a death grip on the rope. I panicked for a few seconds but eventually got into the groove of how to repel and slid down the slope. I could have fallen, but held tight, listened to the instructor, and bounced on down the cliff. And qualified to become a "jungle expert."

Most of the remaining training was how to live in the jungle. We moved out of the field, into the barracks and onto a real bunk bed. Saturday evening there was time for relaxation and the opportunity to visit Colon, the closest city to Fort Sherman. It was time to throw down a few rums and Cokes.

To further acclimate to jungle life, we had a tasting. A sampling of edibles was spread out on a table before us. We passed by and took a sample of each, not knowing what we ate until we had completed the sampling. We kept thinking we were given monkey meat, but it was some sort of fish or lizard. There was a variety of fruits to sample, to include plantains, coconut meat, and pears. For one of our meals, we were presented with a live chicken and were required to prepare it for consumption. Fortunately, a member of our group was a former farmer and was familiar with how to dress our chicken.

The last big training mission was a nighttime escape and evasion exercise. The group was dropped a distance from the objective and told to walk back. The whole time we were to avoid capture by the enemy, and if you were captured, you would be tortured. A hazard of nighttime maneuvering in the Panamanian jungle is the "black palm." This tree has a slender trunk with long, very sharp spike needles surrounding it. If you met one of these trees in the dark night, you would feel its wrath.

My group managed to get through without capture, but that was not the end. When we got to what we thought was the objective, we hit another obstacle. To finish the exercise, we had to navigate across a lake. Of course, we did not have a boat or raft, however, we came across a native, who was part of the exercise, who would sell us one. The native only spoke Spanish, but he was willing to trade cigarettes for a raft. We struck up a deal, got a raft, and paddled our way to safety, ate a hot meal, and took a much-needed shower.

This was the finale of "Jungle School." We departed to Fort Riley, Kansas and were well satisfied that we were now ready for the jungles of Vietnam.

However, the adventure continued as the plane flying us back to Fort Riley became low on fuel. It was late on a Saturday evening, and the pilot could not find an airport open that had fueling facilities. We flew around for a bit and finally found available fuel at Paducah, Kentucky. We landed, gassed up, and proceeded to Manhattan Airport, ending our two-week adventure.

(Editor's note from Bob Babcock: In February 1966, I was given a chance to attend jungle school. My logic was… I don't need to train to be miserable, and why would I leave my wife for the 2–3 weeks of the class when I was heading to Vietnam for a year. I passed it up and got along like all our other troops in the jungles of Vietnam.)

CHARLES "CHUCK" BAYLESS

BRANCH AND JOB: US Army - Armor Officer, UH-1 Helicopter Pilot
DATES OF MILITARY SERVICE: November 1964 through May 1968
DATES SERVED IN VIETNAM: July 1966 to July 1967
UNITS SERVED IN VIETNAM: July 1966 to Jan 1967 Troop C, 1/9th Recon Squadron, 1st Cav Div: Jan 1967 to July 1967 48th Assault Helicopter Co
HIGHEST RANK HELD: Captain (O-3)
PLACE OF BIRTH: Cleveland, Ohio — 1942

A VETERAN'S PRAYER

Lord, thank you for this beautiful day, for we Vietnam Veterans to come together to celebrate our brotherhood. Among men our age, we are so fortunate to have this time for camaraderie and fellowship, for a purpose. It honors the time and sacrifices we made when we were young.

So often as I stand with this group, Lord, I feel your presence. You have your hand on one man's shoulder, and you are smiling at another. We talk of our ordeals in rice paddies, jungle canopies, Delta waters, foggy swamps, and the cloud shrouded Central

Highland mountaintops. You smile at us and say, "I know, I saw it all, I was there with you."

Lord, give us the wisdom to understand why we survived the Vietnam War and many of our buddies did not. These are questions that we have to this day, and we welcome your insight that can work like a balm on our thoughts and emotions. Help us be strong and reach peace with this.

From our experiences in Vietnam, we can take to heart the words of author Tom Purcell, "Each veteran who gave the ultimate sacrifice represents a life cut short—young men and women who never came home, never started a family, never got to see their freedoms they died to protect. Their sacrifice is not just national—it is heartbreaking personal."

Help us to understand that we did accomplish a great saving effort for the South Vietnamese people. We kept Communism from their doorstep during those years. We were not miracle workers; we could not protect them forever. Bless them, Lord, for thankfully, today, they show us gratitude for our herculean efforts to stave off that evil force.

Upon our return from Vietnam, the flag meant so much more to us, Lord. We like the way Tom Purcell's father put it—"The flag should be one of those reminders—of principles: freedom, responsibility, shared purpose, and a continuous desire to improve. For some, our flag can be a symbol of our constant striving. America is exceptional because its people are free to speak, worship, create, and build as they pursue their own happiness. Preserving our freedoms requires vigilance and respect from every generation."

Lord, help us seek out opportunities to educate today's young families about the existence of war, that there will always be conflict where one faction wants to crush another. We cannot sugar coat it. So, help us, help them, realize we need our strong military. We need courageous, country-loving, God-fearing men and women to step forward to protect our nation. We have a responsibility to spread this message. Please give us the wisdom to do it with honor and dignity.

Bless our military brothers and sisters, those on duty now, those who have served and face health issues, and those who — after selfless service — have gone to the Great Beyond.

Thank you for the goodness and mercy we receive every day. May we enjoy this time with our brothers, knowing that it was You who brought us together. In turn, may we strive to do good works in your name. Let us not only hear your voice, let us obey it.

AMEN

NOMIKOS "MIKE" PATELLIS

BRANCH OF MILITARY AND JOB HELD: Army, Field Artillery, XO of Field Artillery Group.
DATES OF MILITARY SERVICE: 1958 to 1979
UNITS SERVED WITH IN VIETNAM: Advisory Team 59, Mekong Delta (First Tour) XO of 1st Field Force Artillery Group (Second Tour)
DATES SERVED IN VIETNAM: Nov 1965 – Oct 1966; Sep 1970 – Aug 1971
HIGHEST RANK HELD: Lt. Colonel (O-5)
PLACE AND YEAR OF BIRTH: Greensboro, NC, 1938

FIESTA-AT-NIGHT

My first assignment to Vietnam was 1965-1966 as the Intelligence Advisor for MACV Advisory Team 59, with duty station in Ca Mau, some 250 miles southeast of Saigon in the Mekong Delta, adjacent to the U Mien/ Nan Cam Forest. Upon arrival, I was quickly made aware that this triple canopy forested area was the primary Viet Cong (VC) training and R & R area.

Because of the vulnerability of our geographical location, we accepted an offer to reside in several vacant buildings on the grounds of the Province Chief. Concurrently residing with us was a small US Air Force flight control center which managed the air space of the entire ARVN 4th Region.

The local ARVN soldiers, militia, and civilians were continually being killed and injured by buried explosive devices along streets, roadways, dikes, and occasionally submerged in canals. The majority of these devices were unexploded ordinance consisting of rockets, mortar shells, artillery rounds, and bombs.

My Vietnamese counterpart and I devised a plan which was approved by the ARVN Corp commander and his senior US Air Force commander.

The plan incorporated a small number of 12 hour Delayed Fused 250- and 500-pound bombs to be included in several local B-52 sorties scheduled for our area. Hopefully the VC would collect and move this unexploded ordinance to their work areas.

Several weeks passed and we received a message from our Flight Control Center that two Rolling Thunder sorties were scheduled for our Area of Operations (AO) with dates and coordinates to be provided later. A week later, I received the missing data in a classified message.

Two weeks passed and at 0900 hours on the given date, we heard the thunder and felt a slight ground vibration. At 2000 hours, we launched two OV-10 aircraft w/observers to observe the target area for signs of explosions or fires. Over the next two hours, two large and three small explosions were reported. One observer reported his area looked like a Fireworks Fiesta from 5,000 feet.

The following morning staff briefing became known as the Fiesta-At-Night briefing. Several additional delayed detonation missions were conducted, each with similar results. The number of large indiscriminate explosions sharply decreased as the observed explosions were determined to have taken place within VC workshops where the ammunition had been transported to be reconfigured.

ROBERT O. "BOB" BABCOCK

FIRST WEEK IN COUNTRY

6 August 1966: The LST (Landing Ship, Troop) moved alongside the USNS Nelson M. Walker, waiting to take another load of troops ashore. As I loaded the LST, my mind wandered to the other times American troops had boarded LST's to land on hostile shores.

This time, we did not have to climb down nets to get into the boat, and we did not have to worry about fighting our way ashore. All we had to do was walk out a side door of the Walker, carrying our rucksacks and duffel bags, and stand tightly packed on the LST while we waited for it to finish loading troops.

I much preferred our way of landing to that of our World War II Fourth Infantry Division counterparts. It was D-Day, June 6, 1944, exactly twenty-two years and two months earlier. They had stormed ashore on Utah Beach through a hail of German gunfire, not knowing if they would survive the day... many did not.

I was very uncomfortable though. They still had not given us any ammunition for our rifles. They could tell me the area was secure, but I did not like being unarmed.

At the airfield, a crusty Air Force sergeant told us to line up, go by the tent they had set up to issue ammunition, and pick up four boxes (80 rounds) of M-16 ammo each. He didn't carry a weapon and seemed amused at the uneasiness of this new bunch of green American troops.

It is a well-known fact that American soldiers hate to stand in line. That was not the case here. As soon as we got our 80 rounds of ammo, most of us went to the end of the line and waited to get more. Before they broke the line up, I had 160 rounds in my ammo pouches. Most of my men had 240 or 320 rounds and were still standing in line for more.

Soon we got the order to load onto a C-130 airplane for the trip to the central highlands and our base camp at Pleiku. Having jumped from a C-130 in Airborne School, with 60 of us on the plane, and with all the extra equipment we had, I guessed we would get no more than 50 on the plane. They loaded all 180 of us onto that C-130. Safety rules in a combat zone are different than stateside.

We were too crammed in to sit on our duffel bags, we all stood, now knowing what sardines must feel like. As I looked over the sweat drenched faces around me, there were feeble attempts at humor, but most men just looked straight ahead, wondering what was next. As the plane rumbled off the runway and gained altitude, we welcomed the coolness.

The flight to Pleiku lasted only half an hour. Lieutenant Fiacco rode in the cockpit with the pilots. Excitement oozed from him as he jumped down from the cockpit. He had watched as one of our Air Force fighter planes blasted the ground with napalm in support of troops engaged with the VC.

This story only heightened our concern that this was the real thing, and we could probably expect to be fighting before the day was over.

Engineer dump trucks awaited us at the airstrip. We loaded onto the trucks for the trip to our new base camp south of Pleiku. Just as we had been trained, we sat in the truck, facing out, weapons loaded and at the ready, poised for anything that might come our way.

We were again confused about what this place was all about. Driving through the streets of Pleiku, we saw American soldiers walking the streets without a rifle or any of their combat gear. As far as we knew at that time, everyone and everything in Vietnam was potentially bad and we had to always be ready. All these soldiers seemed to be interested in was shouting jeers at this convoy of green "cherries," and going into the next bar.

We had not yet learned that Pleiku was a secure town, and we did not have much to worry about there.

We did not let our guard down as we left Pleiku, heading south to what was to become the new base camp and home of the 4th Infantry Division. It was after 6:00 PM as we pulled into what was to become a sprawling base camp. Now, only a few tents spotted the cleared area which stretched between the thick jungle obscuring the eastern horizon and two tall mountains to the west.

Major High, our battalion executive officer, assembled the officers and walked us to the portion of the perimeter we were responsible for defending. He told us, "The First Cavalry Division spotted a battalion of North Vietnamese troops day before yesterday just beyond that hill you can see to the southeast." He pointed at a hill approximately three miles away. "Now go ahead and get set up."

Keeping everyone alert was not a problem as we got ready to do for real what we had been training for the past seven months. So started my tour in Vietnam.

That first night was described in a letter written to my wife… Our fields of fire consisted of a path the width of a bulldozer blade that had been cut in front of our positions. On the other side of the bulldozed path was dense jungle. We dug hasty prone shelters, loaded our weapons, fixed bayonets, and got ready for anything to happen. Well, it did!!!

It rained, and it rained, and it rained some more. I learned my first lesson in Vietnam — never try to sleep in a foxhole when it is raining. A foxhole collects water, and you will get very wet. All I had to keep me dry was my poncho... it did a terrible job. Every time I moved, a stream of water rushed under the poncho and soaked me. I tried to lay on my gas mask and equipment to keep out of the water. My poncho kept some of the rain off, but I was wet and miserable all night long. It is hard to know what wet is until you spend a night in a monsoon.

I really surprised myself because I was not scared. I was very alert but not scared. A loaded rifle with a bayonet on the end of it makes a pretty good sleeping companion under these circumstances. (This was the last time we fixed bayonets at night for the remainder of the tour, but it seemed like a real good idea that first night.)

As they say, it is always darkest before the dawn and that is true in Vietnam. That is also the time you must be the most alert to attack. We were. When it finally got daylight, I was tickled to death to get up and move around. I was wet, cold, and miserable, but everyone was in the same situation. We laughed at each other and maintained high spirits. We all looked like drowned rats.

My next letter, written three days later, continues to describe that first week... The area is changing by the hour. The engineers are busy clearing the jungle with bulldozers. There are about 200 Montagnard tribesmen cutting bamboo and clearing the area. These people are primitive. They wear nothing but a loin cloth and a ragged old shirt.

We have killed a couple of bamboo viper snakes. Bamboo vipers are very poisonous and are the same color as the bamboo they live in. A Montagnard can kill a snake instantly with his machete.

We are not quite as adept as they are—we tend to shoot them and not get close enough to use a machete. We also must be careful we do not shoot one of our troops in the process.

Morale is high among the troops, even though we are all muddy and filthy. Conditions improve daily. We have had nothing to eat but C-rations, but they expect to have a hot meal for us soon.

A Vietnamese madam drove up to the perimeter in a three wheeled French made Lambretta Sunday afternoon and tried to convince me to let my men partake of the prostitutes she had brought out with her.

I do not know Vietnamese, but I have learned one phrase, "dee-dee" which is supposed to mean, "Get out of here!" When I used it on her, it turned her smile into a frown, but she still did not leave. Some of my men were not helping the problem.

When I finally pointed my rifle at her, she got the hint and moved on to another part of the perimeter to see if she had better luck with someone else's troops who did not have such a "bad assed" platoon leader. Who knows what other kind of challenges I will get.

Yesterday was a very bad day. At mid-morning, there was an accident in Charlie Company. A guy was cleaning his pistol, and it went off and killed the guy in the tent with him. It was strictly a case of negligence on the part of the man who fired the weapon, but the repercussions were terrible.

Colonel Miller wanted to make an example of the case, so he relieved the company commander, a West Point graduate. He also relieved the platoon leader and took a stripe away from the platoon sergeant and squad leader. He charged them all with neglect of duty. The guy who shot him is going to get court martialed for manslaughter.

One good thing about it though, it deeply embedded in everyone's mind that you must be extra careful when everyone is carrying live ammo.

My next three letters rounded out the happenings for the first week in country…

10 Aug 66: So far, the mud and bugs have been the most disagreeable things. We still have not fired a shot or been shot at.

This is really a funny war, or at least up to now it has been. Right now, it just does not seem like we are in a war zone. It is more like one big, hilarious field problem. Everything that happens seems to be different than I expected. Instead of being afraid of the VC, I am more concerned with keeping the kids out of my area. If I look mean enough and say, "Dee-dee," loud enough, they usually get the idea and leave my area, at least until I turn my back and start doing something else.

We are still eating nothing but C-rations. They opened a PX yesterday afternoon. I bought two cases of beer and two cases of cokes to give the people in my platoon. The beer and cokes were warm, but it was a welcome change from nothing but water. We limit each man to two cans of beer a day while we are on the perimeter…

14 Aug 66: My platoon went on our first patrol yesterday. This was the first operation of any kind by our company. Our mission was to go on a reconnaissance patrol of about three miles to see if we could find any sign of enemy activity. We left base camp at noon, walking, and returned around 4:30.

I was first briefed on the patrol night before last, so we had plenty of time to prepare. Yesterday morning I got in a helicopter and flew over the area so I would know what type of terrain to expect. The route was mostly open except for a few rice paddies along a creek and a couple of wooded areas.

Two Army photographers went along, taking pictures. Most of them are to be used in the Army historical record of the war. It was interesting to all my people and me to know our pictures were being taken, so we tried to look rough and rugged.

It still rains every night and day, and the conditions are generally miserable. We haven't gotten the hot chow they have been promising us for four days and we continue to eat C-rations, nothing else, for eight days.

Base camp is really getting to be a pain in the ass. Battalion and Brigade are constantly coming up with crap to harass the troops needlessly. They gripe because the people and weapons are not kept clean, but that is impossible. You get dirty five minutes after you clean up. Water is also hard to get at times.

I am really getting used to roughing it. After so long in it, I have learned to just ignore the mud and rain. We now have one of my platoon tents up, so two nights out of three, I get to sleep on a cot in a tent. The other night I have duty on the perimeter and sleep on an air mattress in my hootch. When it rains, I strip my clothes off, take a shower, and hope it doesn't quit raining before I get the soap rinsed off.

Our old leather boots are rotting since there is no way to keep them dry. I understand the jungle fatigues will dry out a lot faster than our regular fatigues—anything will be an improvement.

Really, the conditions are far from ideal, but I am safe and am tough and flexible enough to adjust to them. One month is nearly gone and there are only eleven more to go. The morale of the troops is high, which is important.

JOHN MARK WALKER, JR.

BRANCH OF MILITARY AND JOB HELD: United States Navy, Ship's
Supply Officer

DATES OF MILITARY SERVICE (ACTIVE AND RESERVE): 1963–1992

UNIT SERVED WITH IN VIETNAM: USS Wrangell (AE-12)

DATES SERVED IN VIETNAM: Nov 1965 to May 1966

HIGHEST RANK HELD: Captain (O-6)

PLACE AND DATE OF BIRTH: Memphis, TN – 1941

PROMOTED AS A JUNIOR OFFICER IN THE NAVY

I served on active duty in the U.S. Navy as a Ship's Supply Officer.
The ship was the USS Wrangell (AE-12), a WW II conversion
from a cargo ship to a Navy ammunition ship, which is a specialized version of a cargo ship. Our cargo was 1,000-, 500-, and
250-pound bombs, rockets, and miscellaneous tools to make the
bombs work. We even carried small arms ammo, which was sometimes transferred to destroyers and other small ships, like mine
sweepers.

Wrangell was assigned two supply officers. Both were reserve
officers fulfilling their 3-year active-duty obligation. I arrived in
July 1964, as Assistant Supply Officer, responsible for Disbursing
(Paymaster) and Food Service. The Supply Officer handled stores
(spare parts and supplies), storage of inventory, and Ships Services
like the ship's store, laundry, and barber shop. My Supply Officer's obligated service ended after a year, and I "fleeted up" to take
his place. A new junior supply officer relieved me from disbursing
duties and ship's services. As the senior of the two of us, I chose
to keep responsibility for food service. That worked for a year and
a half, until my obligated service ended, and I left active duty in
1966.

As Supply Officer, I was the junior and the only non-career department head. I reported to the ship's Executive Officer, a Lieutenant Commander (O-4) rank. I was a Lieutenant Junior Grade (O-2) rank. The other three department heads were LDO (Limited Duty Officer) Lieutenants (O-3), who were former senior enlisted men promoted to become officers. Each had more than 20 years' experience.

One time I had put an E-4 (third class petty officer), serving a six-month tour as Mess Deck Master at Arms, on report for continuously disobeying my orders. Such reports had to be submitted to the XO who never took action on it. The disobedient Petty Officer was never disciplined. Another time he embarrassed me in front of the other three department heads during "Eight O'clock Reports!" Had he talked to me in private, I would have cleared things up, and the whole debacle avoided. I ignored his threat and did not receive the punishment.

He continued making my life miserable. I was invited to stay on active duty but would not do so because I did not want to be part of an organization in which a poor leader like that XO could have a negative influence on my career.

I did join a Navy Reserve unit when I left active duty, finding that it was a completely different environment and a positive experience. In March of 1967, after the LT (O-3) selection board met, I was not on the list of new LT's. My Reserve CO asked me if I knew why I was not promoted, and I did not know. He said, "I will check into it."

At the next reserve meeting he said, "I found out why you were not on the recent LT selection list; you were promoted by the board that met last summer. You are already a LT."

I had not been aware of selection boards because my previous promotion from Ens to LTJG seemed automatic. So, Wrangell XO struck once again. He probably knew I was promoted in July, 1966, but did not tell me. My date of rank was later than my active-duty expiration date, so it did not impact my pay. My Reserve CO was impressed that I was already a LT, and I was glad to know that the Navy still "loved" me.

The Navy Reserve program was very good to me, and I was promoted to Navy Captain (O-6) before I retired from the Navy Reserve. My reserve experiences were excellent, and I learned a lot about leadership.

EDWARD A. WOODS, JR.

BRANCH AND JOB: US Navy—Seabees
DATES OF MILITARY SERVICE: 1964—1970
UNIT SERVED WITH IN VIETNAM: MCB 10 (Mobile Construction Battalion)
DATES SERVED IN VIETNAM: June 1966 to Feb 1967; Sept 1967 to Dec 1967
HIGHEST RANK HELD: Petty Officer 2nd Class
PLACE AND DATE OF BIRTH: Atlanta, GA—1942

MY INTRODUCTION TO THE NAVY

My military life started out like most every other young man. In the fall of 1963, I received a letter from 'Uncle Sam' saying something to the effect "We want you!" At that time, I was working for The City of Atlanta in the Water Department as a draftsman and surveyor. In addition, I was taking two five-hour classes for a total of ten hours at night school at Ga. Tech. Twelve hours was considered full time and would keep you out of the draft.

I did not want to go into the Army. Prior to reporting for the draft, you could join the branch of your choice. On the Tech campus there was a Navy-Marine Corps reserve center. I wanted to join the Navy and see the world. I joined the Navy and became part of the Navy Seabees, no ship, no seeing the world. I had no idea what a Seabee was, but I was joining the Navy; how bad could it be? I did have to report to the draft, which was in a building on Ponce De Leon Ave., former location of the Ford Assembly Plant. As I had already joined the Navy, I was able to be excused. Oh happy day!

I reported to the reserve unit in January 1964 with the understanding I would attend monthly meetings for two years, then I would go on active duty for two years, then when I returned, I would attend monthly meetings for two more years for a total obligation of six years. This was great, Vietnam was not an issue in 1964.

In June 1964, I rode the train to boot camp at Great Lakes, Illinois. Since I was already in the active reserve, I only had to attend boot camp for two weeks. The DI made me the Master at Arms, I guess because I was the only one with some college. My job was to simply make sure that all bunks were made correctly, the shower was clean, and make sure all stayed awake during class instruction and report names of those that chose not to follow instructions.

I remember in the mornings marching to the chow hall going past the brig and hearing all the shouting of instructions to the prisoners. This would not be a place I would ever want to go to.

We marched everywhere. We got up one morning, about 0200 just to march in the rain on the grinder. It all seemed silly at the time, but later, the intention was to teach discipline. Without discipline you will fail. At the end of two weeks, we got on the train and returned home. In that group was a young man named Larry Ray Riddle. Larry was killed March 1967 in Vietnam on the same road I had been on many times before. He is buried in the Marietta National Cemetery.

April 1965, Carol and I were married. In the summer of 1965, my reserve unit did two weeks training at Little Creek Naval Base in Virginia. Little Creek is the East Coast base of the Navy Seals. I don't remember much other than I had kitchen duty; more discipline. I do remember early one morning having to get up, get in formation, and report to a waiting ship. The ship took us out off the coast of Virginia. We deboarded the ship off the side, down rope ladders into a beach landing craft. We then made an assault landing on the beach. I thought what hell I have gotten myself into. This was not my idea of seeing the world.

In January 1966, I went on active duty for 24 months, 13 of which was spent in Vietnam for two separate tours. My active-duty years are talked about in volume one of *I'm Ready to Talk*.

After active duty I returned home, completed my two more years of reserve time, and continued my life as if Vietnam had not happened. I returned to the City of Atlanta Water Department for a short stay. I went to work for Ford Motor Co. and retired after 31 years. Vietnam was not discussed, and I don't think more than just a few people knew I went to Vietnam.

I have not nor will I ever regret serving my country. I look back and speculate what I would have done had I not received that letter in 1963. Undoubtedly my life changed. I believe all should serve their country in some capacity.

PAT QUIGLEY

BRANCH: U.S. Army
DATES OF SERVICE: 6/3/62 – 12/8/69, USARV Support Command, 101st Airborne Division (Airmobile) Support Command
VIETNAM SERVICE: 4/66 – 4/67 and 1/69 – 12/69
HIGHEST RANK: Major (O-4)
BORN: Atlanta, GA – 1940

HONESTY IS A GREAT RESUME ENHANCEMENT

I landed at Tan Son Nhut Air Base in Saigon in early April 1966. I was a 25- year-old Army Captain who had just left command of a medium truck company in Munich, Germany. The first 48 hours were spent dealing with jet lag and getting used to the heat and humidity.

I then learned I would be assigned to USARV Support Command in charge of the motor pool supporting USARV. It was a complicated unit with about 150 U.S. Army personnel, mostly drivers, and 120 Vietnamese drivers. The vehicle inventory included sedans, jeeps, ¾ tons, stake & platform, 2 ½ ton, 5 ton and 40 passenger buses. In addition to sedans for general officers, 15 or so were assigned as a radio taxi service in downtown Saigon.

Fortunately, I didn't meet the officer I was replacing until an hour before he was leaving country. I showed up at the motor pool where I was introduced to several NCO's and the Maintenance Officer, a W3. They told me my predecessor wasn't feeling well but would be at the departure terminal in a few hours. After a brief tour and meeting more personnel, about four of us drove to the Tan Son Nhut departure terminal where I met the Transportation captain I was replacing. He was hungover, with one arm in a sling.

For some reason, he wanted me to see a 4" diameter roll of bills and ensure I met the owner of Wing Phong garage in Saigon. He mentioned an apartment, tennis courts, and what a great host the Chinese owner of Wing Phong had been.

Two days later, after he had safely landed stateside, I discovered one of the radio taxi sedans under a tarp in the motor pool, damaged beyond repair. The young driver who had been sworn to secrecy until his passenger had departed country explained what happened.

He had driven my predecessor from bar to bar in Saigon and on their late return to the BOQ, they were broadsided by a Vietnamese ambulance. Somehow this young driver wasn't held accountable, as I think my predecessor, though inebriated, talked their way out of the mess, probably because he was leaving in less than 12 hours.

And had I met my predecessor the day before he left, I may have been tempted to accompany him and been in the vehicle that was totaled.

But back to the 4" diameter wad of cash. By only sending vehicles which actually needed brake jobs, transmission overhauls, and engine replacements to Wing Phong garage, I spent $400,000 less than the officer I replaced for vehicle repairs in a 12-month comparison.

Fast forward 3 ½ years. My first and only interview as a civilian was for a job as a traffic supervisor with Avon Products in Pasadena, CA. I got the job and that $400,000 savings was, most likely, very helpful. Honesty is always the best practice.

GEORGE MURRAY

BRANCH OF MILITARY AND JOB: U.S. Army, Armor Branch, Rotary Wing Aviator

DATES OF MILITARY SERVICE: June 1964 – February 1968

UNIT(S) SERVED WITH IN VIETNAM: 335th Assault Helicopter Co

DATES SERVED IN VIETNAM: March 1966 – February 1967

HIGHEST RANK HELD: Captain (O-3)

PLACE AND YEAR OF BIRTH: Grenada, Mississippi, 1942

HOMEWARD BOUND – VIETNAM IN THE REVIEW MIRROR. WHERE'S MY CELLPHONE?

I deployed to Vietnam at the end of February 1966. I reported as a rookie helicopter pilot to the 335th Assault Helicopter Company, which supported the 173d Airborne Division.

Five months earlier, I had enjoyed a wedding with a South Georgia young lady named Tootsie Hickox. When we initially met at Fort Stewart, Georgia, I assured her I was going to Army Flight School, and we would go to Germany together.

She mentioned to me several times that the newspapers were reporting lots of action with helicopters in Vietnam. I simply laughed off her observations and assured her we would be assigned to a unit in Germany. That is what my ROTC instructor told us as we were recruited into the flight program at Mississippi State.

I missed her terribly during this time of deployment. Letters were the most important thing going. I tried to write every day. Mail service was difficult, and often we went several days, or even a week without mail, and suddenly a handful. I wonder what mail service was like when my ancestors went to war in 1860?

We were to receive R&R, aka Rest & Relaxation—seven days of leave with travel. By letter, we planned feverishly to meet in

Honolulu for a honeymoon. Several days before I was to head out to meet my young bride, the 173rd Airborne Brigade, our parent unit, canceled all R&R.

A newspaper writer named Jeannie Dixon predicted the largest battle of the Vietnam War would be fought during the week we were to be in Honolulu. Not sure why the U. S. ARMY thought a newspaper prognosticator would be privy to the plans of the Viet Cong.

No way a letter was going to reach Tootsie before she departed, so my company commander arranged for me to use MARS, a radio telephone system used for rapid communication by high-ranking Army leadership. I placed a very short phone call. We were excited to talk for the first time in six months, but both were disappointed. We were not going to Honolulu.

Such was the status of communication during the Vietnam War. It must've been horrible during earlier wars, but my history of communications story isn't over.

Troops in Vietnam all calculated time in country as one year, or 365 days. We religiously counted those days off and were thrilled when we became known as "short timers." As a "short timer," I had my biggest scare with a night engine hiccup, coming out of an infantry night position in War Zone D. Obviously, I lived through the incident, but scared the devil out of our crew.

Tootsie and I had communicated by letters as to my return, date better known as DEROS (Date of estimated return from overseas). That set date was February 28, 1967. I expected to arrive about a day later in Brunswick, Georgia and be greeted by a happy young wife.

Sometimes guys in our unit got orders changing the date of return by several days. Sure enough, my orders came for departure on February 24. Great... I'm on the way home, four days early. I sent a letter to Tootsie, which arrived a week too late.

No cell phones, very few people used voicemail on their home phones. I jumped on a flight from Bien Hoa Air Force Base, leaving behind a lot of good friends and a lot of memories. As we made our way toward the United States, we made refueling stops and I was able to place phone calls to Tootsie's parent's home to let her know I would be there early.

NO ANSWER, NO VOICE MAIL, NO CELL PHONE. Calls from San Francisco, Dallas, and Jacksonville—all quiet—NO ANSWER, NO VOICEMAIL, NO CELL PHONE.

I rode a Greyhound Bus from Jacksonville to Brunswick. I hopped off the bus, made a phone call, NO ANSWER—NO VOICE MAIL—NO CELL PHONE. I stood in front of the Brunswick, Georgia bus station wondering whether to walk the five miles to Tootsie's house or just wait and make another phone call. Suddenly, I observed a yellow Volkswagen convertible driving past the bus station, driven by a beautiful blond.

I was not a dignified U.S. Army Captain, with coveted Aviator Wings, running down the middle of the street, waving arms and yelling after a cute blond. BETTER THAN HAVING A CELL-PHONE!

I finally got the attention of Tootsie, who was obviously surprised to see me. She was mad as a South Georgia sand gnat that I came home early without letting her know. We quickly cancelled her beauty shop appointment and laughed about the unusual reunion in the middle of Newcastle Street in Brunswick, GA. NO ANSWER—NO VOICE MAIL—NO CELL PHONE

BRUCE BURGEE GEIBEL

BRANCH OF SERVICE: U.S. Navy — Seabees Enlisted and Civil Engineer Corps Officer
DATES OF MILITARY SERVICE: 1960 – 1991 (USAF ROTC, Navy Reserve and Active Duty)
UNIT SERVED WITH IN VIETNAM: Naval Mobile Construction Battalion ELEVEN, (NMCB-11) — Vietnam 1966 and 1968 — Thailand 1967
HIGHEST RANK HELD: Navy Captain (O-6)
PLACE OF BIRTH AND YEAR: Washington, DC — 1942

SERVICE TIME: WHO I WAS AND WHAT I DID, 1960 – 1991

I was born in Washington, DC, on August 18, 1942. At that time my father, Dr. Frank B. C. Geibel, was a doctor (OBGYN) and Commander in the Medical Corps, U.S. Navy Reserve and had served some time on Naval ships in the Caribbean Sea visiting various islands and Honduras in South America.

I grew up in Columbia and graduated from Dreher High School in 1960. I attended college and later graduated with a Bachelor of Architecture degree from Clemson University in May 1965.

After serving two years of USAF ROTC at Clemson (1960-62) I joined the Navy Seabee Reserve Unit 6-3, of NMCB-6, in the Navy and Marine Corps Reserve Center, Greenville, SC, as an enlisted member (1962-1965). I think my knowledge of my dad's service time with the Seabees in the Philippines during WW II was a primary reason I joined the Seabee reserve during my college days. I attended Boot Camp at Great Lakes, IL, in 1963 and Seabee reserve duty at the Seabee base in Davisville, RI in the summer of 1964.

After attending Civil Engineer Corps Officer School (CE-COS) in Port Hueneme, CA, from March to June 1966, I was assigned my first duty station with the Seabees of Naval Mobile Construction Battalion ELEVEN (NMCB-11) then deployed to Da Nang, Vietnam. I arrived in country on June 26, 1966. I suddenly realized how hot and humid the climate was in Vietnam in the summer months. The smell of P-tubes, open sewage, and fire pits where we burned trash, food stuff and other supplies was another thing I remember. I was dripping wet after a 30-minute ride from the Da Nang airport to our Seabee camp in Da Nang East.

My battalion was located across the road from Marine Air Group 16 (MAG-16, Marble Mountain Airfield) and next to the NSA Field Hospital in the Marble Mountain Military Complex. We were all located between Marble Mountain and Monkey Mountain along the South China Sea.

Vietnam - Under Fire: Oh, man—I knew this might happen during my service time in Vietnam, but it was still a surprise and not expected during the first month I was on board with my Seabee battalion. Bingo, Bango, Bongo! Many of us were playing Bingo at our enlisted men's club—The Green Shack—on the evening of July 23, 1966. Suddenly, we heard the Bango—incoming 82 mm mortars fired into our camp site and neighboring military compounds with sounds like someone was beating on a Bongo drum. We were trained during peacetime and field preparations when under fire to find a foxhole or closest bunker and lay low until the Bango stopped. When it stopped, we were to head to our duty station or company command bunker to receive further directions.

I headed to the headquarters bunker since I was assigned as the Assistant Operations Officer (S-3A) and head of the design and surveying crews. I had no other real assignment during blast time, so I stayed in the bunker observing all the chit-chat and seeing what the command staff was doing until the "all-clear" was given about an hour later.

We had five Seabees who received shrapnel wounds with four medevacked to medical facilities outside of Vietnam. All later received Purple Hearts for their injuries by the incoming rounds. We also had some buildings, infrastructure and command vehicles damaged as well.

My impression: *"Holy Sh*t! Stuff like I've always read about during wartime really does happen. And I was part of this happening."*

We finished our first overseas tour in Vietnam between October 9-18, 1966, and our battalion headed home on various aircraft to Naval Air Station (NAS) Point Mugu, CA. After a period of up to 30 days leave, we would then undergo another 3+ months of classroom and field activities, including pistol and rifle firing on the firing range along the Pacific Coast below NAS Point Mugu and then two weeks of intensive military training down at Camp Pendleton, US Marine Corps Base, in southern California before being deployed again to Vietnam.

In the meantime, I was assigned as the Assistant Delta Company Commander (one of two construction companies) and slated to be the Delta Company Commander during the next deployment. But, due to an emerging situation, I was subsequently assigned to be the Officer-in-Charge (OIC) of a 13-man Seabee Team scheduled to deploy to Chiang Kham, Thailand in 1967. We would be working with a 16-man Thai Border Patrol Police (BPP) team under the guidance of the CIA along the Thailand–Laos border, just across the Mekong River. We were transported from our site to remote Hill Tribe villages on Air America aircraft. The goal was to perform civic action projects in various Hill Tribe communities along the border with Laos and help keep the North Vietnamese and Viet Cong (VC) enemy forces from coming over the border into Thailand. We had a Chief Petty Officer assigned as the Assistant OIC of the team.

After 12-16 weeks of training where all team members were crossed-trained in the seven Seabee ratings—builder, equipment operator, equipment mechanic, electrician, steelworker, plumber, and engineering aid—plus the medic (Corpsman) assigned to the team. Each team member knew the "how's and why's" of each of the other team members—i.e., where the basic techniques in design, building, equipment operating, surveying, equipment maintenance, steel-working, well-drilling, and first aid were stressed. Each team member could fill in various situations where more manpower was needed.

This training culminated in a two-week field exercise at a remote location to enhance our teamwork and skills. We went to the U.S.A.F. base at Cape Vandenburg, CA, to work on roads, drainage, and other projects to enhance our self- preservation, skills, and teamwork. We had to work as a team and not as individuals with such close living conditions and working accommodations at the overseas site where we were in Thailand.

We did our six-month tour in Thailand as one of three such teams sent over when we went—teams from my NMCB-11 battalion as well as similar teams from NMCB-9 and NMCB-10 all out of Port Hueneme, CA. We were assigned to three different Thai village sites along the border with Laos.

A summary of my Seabee Team 1109 activity concluded we had carried out a very rigorous Civic Action Program (CAP). Project work included construction of 32 timber bridges, 12 hand-dug wells, six earthen dams, work on 22 schools and 12 Buddhist temples even if it was only to provide construction advice or much needed materials they couldn't purchase locally, four short take-off-and-landing (STOL) airstrips and improvement of over 43 miles of roads in northern Thailand.

In terms of how the team was able to reach the people, the Medical Aid Program (MAP) through Corpsman (HM1) Richard Lee (Doc) Brown, and his BPP counterpart medic treated some 13,000 patients in 110 different villages during our deployment.

After that tour of duty, I was again back in Port Hueneme, CA, for more training in preparation for another tour of duty in Vietnam. I was made the Assistant Delta Company Commander since I was only going to be in Nam for about four months before rotation to a follow-on tour.

We deployed to Quang Tri, Vietnam in the northern I-Corps, just south of Dong Ha and the DMZ between May 9-14, 1968. I arrived on site on May 15, 1968, on the next to last flight into Da Nang. This was shortly after the NVA and Viet Cong (VC) had pulled off the TET Offensive starting in February 1968. It was pretty much over with when we journeyed into Vietnam in May.

During this deployment, we had a large roadwork project that started so we could begin around-the-clock delivery on 24-7 shift work seven days a week. We built aircraft revetments, lengthened, and installed Marston Matting for runways and landing pads in Quang Tri and built a new Ammunition Supply Point (ASP) in Quang Tri to get ammunition, bombs and other ordinance relocated out of Dong Ha to Quang Tri. It had to be relocated since the ASP in Dong Ha was in rocket and mortar range of the NVA and VC enemy forces who'd blown up that ASP several times, resulting in deaths and injuries to Seabees, USMC, and other military personnel stationed in Dong Ha.

We came under attack several times with rockets fired at our camp site by the NVA and Viet Cong during our deployment. We saw the smoke from the ASP in Dong Ha being hit again with rockets and mortars and felt the shockwaves from the explosions all the way down in Quang Tri. I was then taken out of Delta

Company and reassigned to Alpha Company as Project Manager for road work and to barge large rocks up from Da Nang through the Cua Viet River to our rock crusher plant. Here it was crushed into small fines for concrete mixture and larger crushed rock for laydown on Vietnam Route One from Dong Ha to Quang Tri. Rollers would pack the small rock into the roadway prior to paving certain sections for a smoother run for construction vehicles between those two cities.

We had Seabees injured and trucks damaged or destroyed by mines set in the road by enemy forces and periodic small arms fired at our vehicles, especially out on Vietnam Route 9. It was quite a project, and we eventually began the upgrade of Vietnam Route 9 from Dong Ha to the Rock Pile and Camp Carroll on the way to Khe Sahn.

During this tour of duty, I took R&R back to Honolulu, Hawaii, for 10 days and married my sweetheart, Sandra Mary Fowler, on August 23, 1968, in the Hickam USAF Base Chapel. I then returned to Quang Tri where later that month I was offered a commission in the United States Navy and gave up that old "Reserve" title.

During my tours in Vietnam and Thailand, I earned the following medals and ribbons for my wartime duties and responsibilities: *Combat Action Ribbon*; two *Meritorious Unit Commendations*; the *Vietnam Service Medal* with fleet Marine Corps distinguishing device and five military campaign battle stars; *Vietnam Campaign Medal* with 1964-device; *Expert Rifle Medal*; *Sharpshooter Pistol Ribbon*; and other medals and service ribbons while serving in NMCB-11. Welcome Home all you Vietnam Veterans.

I returned stateside in September 1968, and my wife and I took off for Norfolk, VA, where I was assigned to the Resident Officer in Charge of Construction (ROICC) Office as an Assistant ROICC for the Naval Air Station, Norfolk, VA. This was a start for the next 24 years of active duty in the Civil Engineer Corps of the United States Navy.

It was at this time that I looked back at my uneventful time in an architecture and engineering firm, where I would basically be a draftsman and specification writer at minimum wages and do no real design or oversee any construction projects for at least two years or more. I would basically be just a draftsman and therefore I made the decision to stay in the Navy for a career.

I never looked back at my civilian work time in the past and just looked forward to the time I could excel at work somewhere else around the world in the future.

I began a career in the U.S. Navy where 14 of my remaining 24 years of active duty would be spent overseas. I look back at that time as the love of my life seeing things around the world that I'd only studied about in architecture during my college days.

In recent years, I have come to realize that the United States Navy has been a very important part of my life from the days that my father, Captain Frank B.C. Geibel, MC, USNR served in the U.S. Navy during WW II and for me, Captain Bruce B. Geibel, CEC, USN, Retired, during the time I served in the U.S. Navy during Vietnam.

For the "I'm Ready to Talk—Vietnam Veterans Preserve their Stories (as told to Robert O. Babcock)," Deeds Publishing, Atlanta, GA. I honor and praise you for your service to our nation. I sincerely appreciate what each of you has done during your military service days, your civilian work after your service, and work with the various Veteran's groups many of us have joined over time. Welcome Home to you Vietnam Veterans. May God Bless you and your families and loved ones who were with you during your life's experiences.

ROBERT O. "BOB" BABCOCK

LIEUTENANT DEXTER, RN

I first saw her early in January 1967 when I went to pay our troops in the Pleiku hospital. As I entered the ward with the malaria patients, she caught my attention. Her green army fatigues could not hide her obviously well-proportioned figure. Her blonde hair and pretty face added a spark to the dull hospital ward filled with American GIs.

As I talked to one of my men, she approached his bed. He introduced me to her as, "My lieutenant." In our short conversation, it was apparent she was a special person who would do everything within her power to help and comfort the GIs under her care.

As I left the ward, my eyes wandered across the room for a last glimpse of Lieutenant Dexter as she made her rounds, pausing to give a smile and word of encouragement to each GI as she passed him. It seemed like forever since I had seen a beautiful American woman.

The memory of the beautiful nurse lingered with me. When I made my next trip to visit the troops in the hospital, I started looking for her as soon as I neared the hospital grounds. I saw nurses, but she was not among them. After an hour of visiting with our troops, I left, disappointed that Lieutenant Dexter had not been there.

The next payday, I made my hospital visit again (we always had men in the hospital with malaria or wounds) and immediately saw her as I walked onto the ward. She was busy working with a patient across the room from where I was paying my men. It was hard to keep my mind on my payday activities as I kept shifting my eyes to watch her work. She had lost none of her beauty or charm since I had last seen her.

I lingered for quite some time after paying the men, idly chatted with them, and finally left when she left the ward. I am sure she was totally unaware I was in the room watching her work.

In mid-March, when Lieutenant Dean, Sergeant Cheatham, Sergeant Muller, and PFC White were wounded, I caught a helicopter from the fire base to the hospital to check on them. All had serious wounds, and I was very worried as I walked into the ward. The first thing I saw was Lieutenant Dexter working on Sergeant Muller. He had taken two bullets through the back and was in critical condition.

It immediately eased my mind when I saw who was caring for him.

I spent the morning talking to the men. Roy Dean asked me to write his wife and explain what had happened, but to be certain to reassure her he was alright. She was expecting a baby at any moment, and he did not want her to be unduly alarmed. An NVA mortar fragment had ripped through his side. He was in stable but still serious condition.

Sergeant Cheatham was his usual bright and chipper self, despite the serious wounds to his thigh and groin area. PFC White's foot wound was not causing him too much pain. Sergeant Muller was in the worst shape. For several days prior to getting wounded, Doug had been ignoring the fever of an increasingly serious case of malaria as he continued to lead his men through the jungle. The bullet wounds, coupled with the malaria, had him in a life-threatening condition.

Lieutenant Dexter continued to add a touch of warmth and professionalism to the room as she meticulously carried out her duties. She took time to stop and chat but never stopped long enough to neglect looking after anyone who needed her attention. I am sure my wounded friends appreciated her as much or more than I did. They maybe were not in shape to appreciate her beauty as much as I, but they had to enjoy the dedication of an American woman's care as they lay there, hurt and scared, so many miles away from home.

All the men were evacuated before I could get back to see them again. Except for Doug Muller, I have not seen or heard from any of them since that day. I met Doug in September 1987 at LaGuardia airport in New York City and learned about his ordeal during his long recovery period. I feel certain the rest survived their wounds, none of their names are on the Wall in Washington. (I have since become good friends with Doug Muller, talked to him on the phone yesterday, and met Roy Dean a couple of times when I visited Fort Benning.)

I saw Lieutenant Dexter one more time. I never did learn her first name, where she was from, or anything about her. She never even knew I existed. All I know about Lieutenant Dexter is she was a beautiful American nurse who provided a spark and a shot in the arm to me and the wounded GIs she cared for.

On Veteran's Day, 1993, a statue honoring American women in the Vietnam War was dedicated. It stands in front of the Vietnam Memorial Wall in Washington, D.C. I walked in that parade and attended that ceremony with Lieutenant Judi Dexter Richtsmeier, her Vietnam vet (and wounded—they met in the Pleiku hospital) husband, and her sister, *another Vietnam nurse.* (That is another story for later in this book).

These women, and others like them, provided real care and comfort to many of the men who died in Vietnam and to the ones who were wounded and survived, as well as those of us who were fortunate enough to not need their care. It was a comfort to all of us to know they would have been there if we had needed them.

COMING HOME — 13 JULY 1967

The thought, the dream, the hope, the driving force that has kept soldiers going in all wars are those two magic words — "Coming Home." From the day you find out you are going to war; you start thinking about coming home. You think about the possibility of not making it, and try as you will, you can't totally drive those thoughts from the back of your mind. But most of all, you dream about how great it is going to be when you come home.

As a boy, I remembered looking at the old copies of *Life, Colliers, Look*, and *Saturday Evening Post* magazines and seeing the welcome home scenes from Times Square, from the ticker tape parades in New York, and the troop ships coming past the Statue of Liberty at the end of World War II.

I also remembered vividly from my boyhood the celebration, the parade, and the patriotic event held at the high school football field when John "Patch" Patton, a local high school football hero, was released from a prisoner of war camp at the end of the Korean War.

Dreams of events such as those danced in my head during those long nights when I fought sleep while peering into the dense jungle listening for approaching NVA troops. Later, while lying on my cot as executive officer, the dreams were just as vivid.

My short timer's calendar kept ticking down until finally the orders came… orders to leave Vietnam.

I was hesitant about getting on a helicopter to fly out to the company, which was operating in the Ia Drang valley. But I wanted to bid farewell to my friends still out there. I boarded the resupply chopper to spend one last day in the field.

I certainly did not want to be one of those stories told about a soldier who was killed on his last days in country. When the final helicopter of the day came into the patrol base, I was more than happy to get on it and head back to base camp.

A myriad of other thoughts flew through my mind. I was struck by the sheer beauty of the Ia Drang valley, the double waterfall that glistened in the sunlight, and thought of the major battle the First Cavalry Division had waged there just as I was getting out of AIRBORNE school in November 1965.

I also thought of the friends I had made in the past year. When we trained together at Fort Lewis, they had been troops entrusted to my care. Now they were real people... friends I shared a common bond and a lifetime of experiences with. Sergeant Roath was not just a crusty platoon sergeant; he was a guy who had helped me bring our entire platoon home alive.

Sergeants Benge, Muller, Cheatham, Burrell, and Baez had also become friends. As squad leaders, they had not just done their job and what I told them to do, we had come together as a team. They did not hesitate to tell me if I was making a dumb decision—fortunately, I had the good sense to listen to them.

As the helicopter settled down into the base camp, I climbed off for the last time, headed for my tent, and never looked back. That part of my life was behind me.

Out-processing was a breeze as I checked in my field gear, signed over all my duties as Executive Officer, and got ready to head out the next morning to catch the "Freedom Bird" out of Pleiku airfield.

There was one little glitch—Wayne Germek, the supply clerk, wanted me to check in my M16 rifle before he would sign my supply clearance form. In no uncertain terms, I explained to Germek that I had only slept one night in Vietnam without my M16 being within arm's reach, and this night was not going to be my second. He could either get up with me long before daylight to check in my rifle or he could clear me now and pick up my rifle next to my cot when he got up. Either way, I was not going to give up my rifle until I left. He understood and decided he would trust me.

Another thing I left by my cot, and which I have regretted ever since, was my jungle boots. Those boots were the most comfortable things I ever wore, and they had the character to show what they had been through. The treads on the soles were worn thin, the leather spoke volumes of the tough terrain they had taken me through, and they had served me well over many miles of walking and could have told many stories. They would have made a nice memento—but I was not thinking about that then. I wish I had brought them home and bronzed them.

The time passed slowly until the C-141 appeared as a speck in the morning sky and taxied to a stop in front of the Pleiku air terminal. We stood impatiently in line waiting for the incoming passengers to offload (we short timers, of course, heckled them) before we could get on our "Freedom Bird."

Since Pleiku airfield was within enemy mortar range, it took no time for the Air Force to load us aboard and head the plane back to Cam Ranh Bay for refueling prior to starting the long flight back across the Pacific Ocean. After the short refueling stop, a mighty roar went up from the troops as the plane again became airborne and crossed the coast of Vietnam, we were finally out of harm's way and on the way home!

It was a long flight. Instead of a civilian airliner with pretty, round eyed stewardesses, we were on an Air Force C-141 cargo plane with a lanky airman in charge of the passenger compartment. The seats were canvas and faced the rear.

The airman came onto the PA system asking which officers wanted to eat on the way home. Naturally, we all raised our hands. He then informed us we were to go to the pay window when we landed in Guam, pay for the two box meals we would have enroute, and get a receipt to verify it. We knew then that we were out of the combat zone and back into the military bureaucracy.

To further attest to being out of the combat zone, when we

landed in Japan, we had to sit for eight hours while a crew was rested to take us the rest of the way across the Pacific. While waiting, several of us went to the officer's club. For the first time in almost a year, we saw real people leading real lives.

Another huge roar engulfed the plane as its wheels touched down at Travis Air Force Base outside of San Francisco. We had made it! The fear that came across me as we sailed past Seattle almost a year earlier was over. I had made it home alive!

All that stood between me and the last leg of my trip was to get an Army physical, get paid, and finish my out-processing. Then, I would be on the way to my wife who would be waiting for me at the Kansas City airport.

Oakland Army Terminal was a bustle of activity as our busses pulled up and unloaded. By now, it had become the primary processing center for returning vets and operated around the clock—with one minor exception. Officers who were processing out of the Army were handled only between the hours of 8:00AM and 5:00PM. I fell into the last category. It was now 8:00PM—another twelve hours wait before I could continue my journey toward home.

After paying for a room in the holding facility, I found the nearest pay phone and placed the long-awaited call home. My wife was already in Kansas City. We had agreed I would let my parents know when I would arrive. She would stay in touch with them to find out when to be at the airport (this was long before cell phones and the internet). Naturally, my parents were ecstatic to hear from me. Their prayers had been answered—I was home, safe and sound.

Finally, my processing was completed. I was a civilian again and boarded a cab to take me to the San Francisco airport. I drank in the sights as we crossed the Bay Bridge. I appreciated them much more than I had a year earlier.

My first dealings with civilians were at the San Francisco air-port and on the plane flying to Kansas City. I was not spit on, heckled, called a baby killer or dope addict, or ridiculed. All of that came later in the war. Instead, I was ignored.

Although I proudly wore my Army uniform, Combat Infan-tryman's Badge, and appropriate ribbons, no one acknowledged me any more than they did a businessman traveling in his normal business day. I expected more than that and had a hollow feeling until we came into our final approach for Kansas City and my excitement overtook me.

My greeting in Kansas City was much better—Phyllis, my wife, was standing by the gate anxiously awaiting the moment she and I thought would never come. Everything was as I expected it would be as we ran to greet each other.

Three final things still stand out in my mind about my coming home experience.

First, I decided it was too hot to drive to Oklahoma in our Mustang with no air conditioning. The fact the Mustang was the soon to be classic 1965 model, that it only had 36,000 miles on it, and that it was almost paid for, did not cross my mind. All I thought about was I had just spent twelve months in the heat of Vietnam, and I was not going to put up with it anymore.

So, we went to a local Chevrolet dealer, picked out the top-of-the-line model sitting on the showroom floor, called the presi-dent of the bank where I had worked as a janitor all through high school (a close family friend), got his approval for a loan, paid for the car, and drove it off the lot. There went the money we had saved over the past twelve months and began my life of having to work to make payments.

I have often said there should be a law that did not allow soldiers to make major purchases (over $500) until they had been home at least thirty days to get acclimated to reality again.

Second, as we listened to the radio while driving home from Kansas City, we heard stories of a major battle going on in the Central Highlands involving the 4th Infantry Division. Knowing my mother would be glued to the TV and getting all the details, my first question of her after our initial greetings was, "What is going on with the 4th Infantry Division?"

Her answer still bothers me today. "I don't know. I quit paying attention after you got back to San Francisco."

To her, I was all that was important in Vietnam. To me, my friends were still fighting in a war I believed in. And, like most Americans, she went on with her life now that she did not have a direct involvement in the war.

The third and last thing that stands out in my mind is the reception I got in my hometown. Having lived in the same small town of Heavener, Oklahoma all my life, I knew most of the people. As I walked down the streets and saw someone I knew, it was not uncommon for a conversation to go something like this—"I haven't seen you in a long time, Bob. Where have you been?" When I responded I had just come back from Vietnam, I got an "Oh…", there was an awkward pause, and the subject was changed to something like, "Sure is hot today…" To say the least, that hurt.

I put off writing this chapter for three years. It has been the most difficult to write. It finally came to me sitting in church the weekend the troops started coming home from Operation Desert Storm in 1991. The last thing I did prior to starting this chapter was watch the Desert Storm POW's return and be greeted by the Secretary of Defense, the Chairman of the Joint Chiefs of Staff, and a thankful crowd of thousands. I shed a lot of tears watching that as well as the receptions the other returning troops got. I do not begrudge them a single bit of their welcome; they deserved it all. We deserved it, too. And it still hurts we did not get it.

Thank God we are showing proper respect to today's Soldiers, and as long as I have life in me, I will be part of the welcome home for all American Soldiers returning from a combat zone.

TET AND MAJOR BATTLES

(1968-1969)

JANUARY 21, 1968: Siege of Khe Sanh began. The isolated Marine outpost was under siege into April 1968. 205 Americans died related to the battle.

JANUARY 30, 1968: Tet Offensive began by the North Vietnamese and Viet Cong. Over 100 military and civilian installations across South Vietnam were struck in coordinated and surprise attacks on the first day of the Lunar New Year. This caused news anchor Walter Cronkite to announce that the US could not win in Vietnam, even though each of the attacks of the Tet offensive resulted in military victories for the South Vietnamese and Americans. The real impact was on American popular opinion at home and the will of American political leaders.

MARCH 31, 1968: President Johnson announces he will not seek re-election. His announcement was, "I shall not seek and I will not accept the nomination of my party for another term as your President."

APRIL 4, 1968: Martin Luther King, Jr. was assassinated in Memphis, TN. Riots erupted in the streets of every major US city. Many Americans thought the country was being torn apart by divisions over race, equality, and the war in Vietnam.

JUNE 5, 1968: Senator Robert Kennedy, presidential candidate, was assassinated in Los Angeles, CA. Brother of slain President John F. Kennedy, Robert was a source of hope that he could start healing our country.

AUGUST 26, 1968: Riots broke out at the Democratic National Convention in Chicago, IL. The National Guard was mobilized the help the Chicago police in quelling the riots and restoring order.

NOVEMBER 5, 1968: Richard Nixon elected President of the United States, running on a platform of withdrawal from Vietnam. His plan was called "Vietnamization" with a plan to equip South Vietnamese military and withdraw American troops.

DECEMBER 31, 1968: American deaths in Vietnam reached 30,000. By all accounts, 1968 was the most tumultuous year in American history. The number of Americans killed in Vietnam had doubled in a single year and the damage to American popular support was something that would not be recovered for the rest of the war.

APRIL 15, 1969: Woodstock… more than 400,000 people gathered in Bethel, NY for the Woodstock Music Festival. It is considered a definitive moment for the counterculture generation.

APRIL 30, 1969: American troop strength in Vietnam peaks at 543,282 in-country.

MAY 10, 1969: Battle of Hamburger Hill in Vietnam's A Shau Valley. With over 70 killed in action, it sparked public debate of America's strategy in the war. Reacting to public and political pressure, GEN Creighton Abrams altered US strategy from "maximum pressure" to "protective reaction" in an effort to lower casualties.

JULY 20, 1969: Neil Armstrong becomes first person to set foot on the moon.

SEPTEMBER 3, 1969: North Vietnam leader Ho Chi Minh dies at age 79.

NOVEMBER 12, 1969: News of My Lai Massacre in South Vietnam reaches the US. The massacre was the mass killing of Vietnam citizens by US Army soldiers. Despite the atrocities, a US helicopter pilot tried to stop the killings and rescued civilians.

SOURCE: www.vvmf.org/VietnamWar/Timeline

GLENN PEYTON CARR

BRANCH OF SERVICE: US Army Aviator
DATES OF MILITARY SERVICE: 1958-1986
UNIT SERVED WITH IN VIETNAM: 213th Assault Support Helicopter Company (Chinook); B Troop 7th Squadron 17th Cavalry Regiment (Air)
DATES SERVED IN VIETNAM: XO & CO 213th May 1967-May 1968; CO B Troop 7/17, XO 52nd Combat Aviation Battalion 1971
HIGHEST RANK HELD: Lt. Colonel (O-5)
PLACE OF BIRTH: Shawnee, OK—1934

TET 1968 – RINGING IN THE NEW YEAR

It was late January 1968 when I was thinking about the Tet new year that the South Vietnamese people celebrated. I had been in the Republic of Viet Nam (RVN) for about seven months, and I thought I had experienced all that one could or should have experienced in RVN. I was secure in my surroundings.

TET of 1968 was coming in a week. I recollected all I had been told about TET celebrations. Most all warring parties celebrated the Tet new year by leaving the field of battle to return to their house or going to a nearby pub to "tie one on".

Contemplating this, I figured that Tet 1968 would be a piece of cake flying R&R type runs. No big deal. I could put myself on the flight board as the #1 aircraft for the day of Tet and get the newest warrant officer for my copilot.

On Tet 1968, Wobble One, the newest warrant officer in the company, and I pre-flighted at dusk. Afterwards we went to evening chow, then back to the bunk house to ring in the Tet new year and then off to bed.

At 02:30 the operations runner awakened us saying, "We have a TAC-E, Sir!" That is tactical emergency—meaning we have got to go quick. Most of us slept in full uniform and boots. It was a fast run to operations. While we put on our chest armor and helmet, the ops sergeant briefed us on the mission and gave us the paperwork containing radio frequencies, units, and their locations.

We arrived at the aircraft which had just been towed to the start-up area. The crew chief and two gunners had done a brief pre-flight and mounted the machine guns. All other crew men had done a formal pre-flight the prior evening. We were ready. I started the aircraft since the warrant was new, and I always liked to talk a new guy through the start procedure before I let him actually start the aircraft. However, there was no time for talking.

Engines running, I called the tower for take-off clearance in which we were granted. We picked up the load at the normal staging area for TAC-E's, and we were off to the recipient's location. After that delivery was completed, I called our ops sergeant to notify him I was coming into Phu Loi and would gas her up and put her to bed. The ops sergeant said, "Ok on the gas, but not to bed! I'll meet you on the refuel pad."

I looked over at my copilot as he asked, "Sir, what is going on?" Looking around and observing my surroundings I replied, "I don't know, but I find it odd that all these farmers are burning off their fields on their Tet celebration."

At the refueling pad, Sgt Bentley came on board with a hand full of new mission orders. I immediately asked the sergeant why the farmers are burning their fields. He replied, "That's not their fields. It's the Province and Hamlet Chiefs' homes. Most of them have been executed. We are in a war all over RVN. Here are your new missions. I'll bring food the next time you refuel here. The Division has ordered us to put up eight more Chinooks for the next twelve hours."

So, my short TAC-E delivery mission extended into a mission

period of about 26 hours. What happened to me ringing in the Tet new year and cooling it in my bunk during a no action period?

LARRY TAYLOR

BRANCH OF MILITARY AND JOB HELD: USMCR, squadron pilot thru Commanding General
DATES OF MILITARY SERVICE: 1959 - 2003
UNIT(S) SERVED WITH IN VIETNAM: Air America
DATES SERVED IN VIETNAM: (Laos) January 1967 - August 1968
HIGHEST RANK HELD: Major General, USMC; Captain, Air America
PLACE AND YEAR OF BIRTH: New York, NY 1941

NATIONAL LAMPOON – TET "VACATION"

At Air America, we got six days off per month. Sometimes, you could take the last six days of one month, and the first six days of the next.

I did that in Jan/Feb 1968, with the intent of visiting my pal, Cary King, in RVN. It was Tet, and there was going to be a "truce" right?

There always had been an unofficial truce at Tet. I had a good plan... I thought. I flew into Saigon commercially and bummed a ride on an Army Huey out to Di An, King's base. He was flying an AO mission when I got there, so I just put my feet up on his desk and waited.

You've never seen a double take like the one when he came back and saw me there.

King managed to talk his boss into a three-day R&R, and we took off in a Caribou next day for Vung Tau, where there had always been an unofficial "truce." Vung Tau was rumored to be an R&R spot for the enemy, also. We had a good plan... we thought.

We checked into the R&R Hotel at Vung Tau. Our plan was to watch the movie ("*A Man For All Seasons*"), then hit the 'ville to do what young guys do on R&R. We had heard rumors all day about firefights breaking out at various places around RVN, but we were on R&R, right?

Halfway through the movie, a firefight broke out several blocks away. The electricity was shut off, and we realized then that we had stumbled into the biggest enemy offensive of the war.

At that point, King started trying to get back to his unit, which was in heavy combat, and I started trying to get back to Air America, where a similar upsurge in enemy activity was taking place in Laos. After all, I was a "tourist," and all commercial air traffic had ceased in RVN.

Here's where our adventures diverge.

I bummed a ride on an Air America C-46 back to Saigon, and began 10 days of

bouncing all over RVN, dodging firefights, Da Nang, Nha Trang, etc., looking for a ride out.

Finally wound up back at the Air America compound in Saigon, sleeping on the floor, and manning the perimeter with a trusty WW2-vintage M1 Carbine from the Air America armory.

There was a lot of fighting nearby, but the most dangerous thing that I experienced during those few days was a firefight between the nervous friendlies, Air America. USAF, VNAF, etc. I don't think anybody was hit.

Finally, after haggling with RVN bureaucrats about my lack of a visa, I ignored the

requirement to depart legally and bummed a ride on an Air America C-47 from Saigon to

Vientiane, Laos.

CARY KING

BRANCH OF SERVICE: U.S. Army — Artillery and Infantry Officer
DATES OF MILITARY SERVICE: 1963 to 1987
UNIT SERVED WITH IN VIETNAM: 2-28 Infantry Regt., 2nd Brigade,
1st Infantry Division: 1-7 Field Artillery Bn.; HHB, Division Artillery: Dates in
Vietnam: April 1967 to October 1968
HIGHEST RANK HELD: Lt. Col. (O-5)
PLACE OF BIRTH AND YEAR: Atlanta, GA — 1941

1968 TET OFFENSIVE AND SUMMER COUNTER OFFENSIVE

In the book *I'm Ready To Talk,* in mentioning many of my experiences in the Tet Offensive of 1968 as the S-2, intelligence officer, of the 2d Brigade of the First Infantry Division in Di An, I left out the best story of all. I should have included the unexpected visit by my best friend since 1954, Larry Taylor, who is also a retired United States Marine Corps reserve Major General.

In January 1968, Larry Taylor was stationed in Thailand with Air America, a civilian organization which was part of the U. S. Central Intelligence Agency. Larry Taylor, at that time, was flying H34 helicopters to support the CIA's mission, in the area where U.S. forces were prohibited from operating.

My base camp was located in Di An, Vietnam. On this particular date, either January 29 or 30, 1968, one of my functions as S-2, the Brigade Intelligence Officer, as a qualified aerial observer, was to fly missions every evening starting around 1800 hours until approximately 2100 hours. My mission was to seek out possible Viet Cong and North Vietnamese Army ("NVA") units, and to direct artillery, gunships, or air strikes to destroy them as priority targets and to gather and to confirm or deny intel.

After each flight, I had to return to Brigade headquarters and be debriefed as to what happened during the flight that evening, to include identifying enemy units spotted in or contacted and any action taken or casualties or KIA's, if any.

In December of 1967 into early January 1968, it had become very quiet, considering we were in a combat zone and not many enemy units had been spotted. That night was such a night. We had had enemy contact and when I finished my debriefing, Bob Duker, the Brigade S-3, Operations Officer, said to me, "There's a civilian guy in your hooch who says he's a friend of yours." Knowing soldiers in combat liked to lighten things up, I assumed it was a bad joke, so I said, "BS" and headed back to my hooch, not expecting to find anybody.

But like the nursery rhyme of the three bears, Larry Taylor was asleep on my cot. I was glad to see him, of course, but couldn't help saying, "What kind of a fool visits Vietnam in the middle of a war?" Unbeknownst to us, Tet 1968 was not to be quiet or a ceasefire but was to be the beginning of a major offensive by the North Vietnamese army and the Viet Cong, attacking over 100 places throughout Vietnam.

That night, Taylor and I went to the officers' club, basically a tent with beer and alcohol, and had a few beers. We decided that because this was a quiet time, that I would try to get a three-day pass from the Brigade commander to go to a resort town on the South China Sea called Vung Tau.

My boss, Colonel Thibodeaux, granted my request, with both of us thinking all that was going to happen was this quiet possible ceasefire so, why not let me go for a couple days with my buddy.

The next morning, Taylor and I hitched aircraft rides from Di An to Bien Hoa and then on to Vung Tau where we arrived sometime in the afternoon of what I believe to be January 30th.

We checked into a resort hotel, had a few beers, caught a few

Z's, and were watching a movie which upon its completion was to be the prelude to going out to the local establishments, where you might expect military people, in the middle of a war, to frequent.

Unfortunately, it was not meant to be. In the middle of the movie, *"A MAN FOR ALL SEASONS,"* the power went out at the hotel and the movie was only half over.

We didn't even get to finish the movie for a period of about 30 years. We also got word that the Viet Cong ("VC") had attacked several locations in or near Vung Tau and we, therefore, were restricted to the hotel. I found this to be laughable since I just left an area where the enemy was within 100 meters, so I could not understand why everybody was panicking about what I thought was a little small, isolated attack.

I argued with the Military Police ("MP's") and tried to pull rank so we could get out to the fun places, but to no avail. So, we had a few more beers and crashed.

I'm a little unclear as to whether the next morning was January 31st or February the 1st in Vung Tau. But by the time I woke up the next morning, the hotel manager advised me that my three-day pass had been cancelled, and I was to report to Brigade headquarters ASAP. All l knew was that the Brigade was in heavy combat, involving all the battalions.

That was the first time I heard that they had also attacked Saigon and other locations such as Bien Hoa and Long Binh.

Based on this, as I left the hotel, I told Taylor, "I've got to leave you, and I can't stay. I'll check you out later, or I'll drop you a line when I find out what's going on, but I'm out of here."

This war and this trip to Vung Tau was to be *no fun*, and this three-day pass was a 12-hour pass, as it turned out.

The Brigade had three infantry battalions under our control and all those were involved in heavy contact. They were waiting for me with my weapon and my gear when I landed at Brigade headquarters, what we referred to as the "Dagger" pad.

The CO told me to get to the Thu Duc water plant because intel was that they were being attacked that night. I was to report to him every hour, or more often if needed. Two companies of the 2nd Battalion 18th Infantry were already there. We fought and won that night.

For the next several days and even weeks, we had multiple locations and were continually on the move, engaging numerous enemy units. Chaos was everywhere. NVA and VC were firing mortars from roof tops in the middle of Saigon as we headed to Tan Son Nhut Airbase in Saigon. Bodies were strewn everywhere. Tan Son Nhut Airbase itself by this time was under attack and we all knew that there had been more than 100 places hit all over the country.

I had long since lost contact with Taylor, but, while I was with one of our units at times enough for a briefing to numerous General officers, we got word that at the civilian end of Tan Son Nhut airfield, a shooting had broken out in the civilian terminal.

I was told to find out what it was all about and report back, and if needed, to send somebody down there to take care of it. Shortly thereafter, we were told it was a "nervous Air America" guy who got scared and fired a shot inside the terminal, but it was OK. The word was that it was just a misunderstanding by some folks who were a little nervous.

It wasn't until many, many years later when Taylor and I were together one night and Vietnam and the Tet offensive came up, when I found out for the first time, that on that evening, we were not more than a kilometer or two apart inside the base since Taylor was at the civilian terminal while I was at the military side.

While we discussed that there had been a shooting that night, Taylor steadfastly asserted it had been some other Air America guy, according to Taylor, who got nervous and fired a shot. He continually denies that it was him but, of course, it's became a great joke between us and I always tell everyone, that it was probably Taylor who fired the weapon that night because he got nervous.

Of course, you know Marines don't want to hear that. In spite of that, he and I remain best friends now for 71 years. All in all, it was a chaotic five or six days after Vung Tau and a chaotic three or four months, but it sure gave us both a lot of great stories about his "tourist visit" to Vietnam in the middle of a war.

AUTHOR'S CAVEAT: Please note that in this story, I never referred to myself as an INTELLIGENT Officer, just as an INTELLIGENCE officer; the distinction is intentional.

DANIEL D. WRIGHT

BRANCH OF MILITARY AND JOB HELD: US Army Armor, CAV Troop XO and Troop Commander
DATES OF MILITARY SERVICE: 1965 - 1970
UNIT(S) SERVED WITH IN VIETNAM: MACV, ¾ Cav 25 Inf Div
DATES SERVED IN VIETNAM: September 1966 - April 1968
HIGHEST RANK HELD: Captain (O3)
PLACE AND YEAR OF BIRTH: West Chester, PA 1943

TET 1968

I've been encouraged to share my Vietnam experience in January and February 1968. This period has been referred to as the Tet Offensive.

In the early morning of 31 Jan 1968, I was awakened by one of our radio operators (I think it was Andrew Giordano) that the unit was under heavy attack near the Ton Son Nhut Air Base. At the time, I was the Executive Officer of C Troop, 3rd Squadron 4th Cavalry, which was assigned to the 25th Infantry Division. The division had been put on high alert as there was sizable movement by the enemy as the Tet observance drew closer.

C Troop was directed to move rapidly from our base in Cu Chi and proceed southeast toward the air base. The troop didn't realize it then, but when it got to the air base and encountered heavy fire, they had driven into a three-battalion enemy assault.

Within five minutes, the Troop Commander, CPT Leo Virant, was seriously wounded and command of the troop fell to the Scout Leader of the third platoon, SSG Gary Brewer. This is where I came into the picture.

Our Squadron Commander, LTC Glenn Otis, radioed back to the Troop HQ and I was told to gather up every operational vehicle and all able-bodied men. We were to proceed immediately to the combat action outside Ton Son Nhut Air Base and give relief to the unit. Troop B was also directed to the action.

When I got there, much of the fighting had tapered off. I saw a field of slain enemy soldiers off to the western side of the highway, too numerous to count. These were North Vietnamese regulars, not the pajama clad Viet Cong. This told us we were involved with a major campaign from the enemy.

LTC Otis put me in charge of C Troop since CPT Virant was wounded and had been med-evacuated from the area. My first mission was to take C Troop and elements of A Troop into Saigon and go downtown to the American Embassy. We were to set up a defense of the embassy.

The detachment from A Troop was sent over to the ambassador's residence and set up a defense there. The amazing thing about all this is that at the beginning of the day, I was Charlie 5, Troop Executive Officer, and at the end of the day, I was Charlie

6, Troop commander. This was heady stuff for someone who had only been in the Army a little over two years.

Here's a little background on my assignments prior to joining Charlie Troop. Following the Armor Basic School, I was originally assigned to the 2nd Armored Division in Fort Hood, Texas. Halfway through the school, I was reassigned to the training center at Fort Benning, Georgia. When I got there, I was assigned to the Reception Center. This irritated the hell out of me. I wanted to be in a combat arms unit, and I spent nine months shepherding brand new recruits from civies into fresh green fatigues and buzz haircuts. I was surrounded by a bunch of lame AG officers. I really wanted out of there.

I got my wish with orders to Vietnam.

I arrived in Long Binh in September 1966, with no idea what I was in for. I ran into a couple of 2LT's I knew from Fort Knox, and we made a few trips to Tudor Street. Everyone was getting assignments except me. After all my buddies got sent out, I went up to HQ to see why I was still there. Apparently, I was being paged by loudspeaker to report to HQ, but I never heard the page. It seemed that MACV was looking for a press officer, and I had a degree in English. So next thing I know is I am on my way to MACV.

I got there and showed them the transfer orders, and I never heard a word about being a press officer. I ended up in the city of Dalat in the central highlands of Vietnam. This was one of the safest places in the whole country. I was assigned to a team providing advice to the local militia and the military government. This was a lot more interesting than the Reception Center at Fort Benning. However, it still was not what I was looking for.

Towards the end of my year with MACV, I realized I had never been in a line unit and never had any troop duty. I was disappointed, to say the least. So, I elected to extend my service time and requested to be assigned to a regular combat unit.

After a 30-day leave where I went home after the year with MACV, I arrived at the HQ of the 25th Infantry Division. They were expecting me as my request to be assigned to a combat unit had reached them. The 25th HQ had forwarded my request to the 3rd Squadron, 4th Cavalry. Major Johnson, the squadron S-3, had replied to HQ they could use me as attrition would always make room for a junior officer, regardless of experience.

Initially I was assigned to operations as all the officer slots in the line troops were filled. I didn't stay there long.

Charlie 6 Captain Leo B. Virant was not happy with his XO and wanted him replaced. I'm not sure he got what he wanted, but I got the gig. At last, after two years of crappy assignments, I was in a line unit. I loved being the XO of Charlie Troop. Best troopers ever. It was a little awkward at first, but I soon felt good about what I was doing. I got to know most of the men.

I spent most of my time in the base camp as I was responsible for the mess hall, supply tent, and the motor pool. Vehicle maintenance in an armored unit is huge. Leo would let me go out in the field several times when the situation called for it. On one occasion, he had me lead the troop on convoy duty from Go Dau Ha back to Cu Chi.

All this background information is to let folks know that I was the least experienced dude to ever end up as CO of C Troop 3rd Squadron, 4th Cavalry, 25th Infantry Division. But folks, that happened, and I did the very best I could under very difficult circumstances.

Now back to my Tet story. We stayed at the embassy for two nights and then returned to the squadron that was still in the Ton Son Nhut Air Base area. All the squadron was together and basically for a couple of days just mopped up the area.

On 6 February 1968, the day started out OK, then we got word a company from the 1/27th Infantry was under heavy fire and pinned down in the village of Hoc Mon. LTC Otis ordered Charlie Troop to go in and pull the Wolfhounds out.

I had been a captain for a very short time and CO of Charlie Troop for a week. I'd never been in combat, and my first experience was in this little village north of Saigon. You have no idea. Hoc Mon was awful. I knew things were dicey when the locals were high tailing it out of there and heading for Highway 1.

As soon as we got to where the infantry company was bogged down, we came under heavy fire. We took heavy casualties, including the Third Platoon leader, two staff sergeants, and three other enlisted men. Except for one, all were under age 25. This was a bad day for Charlie Troop. We also lost a tank. I got into a little trouble with losing tanks. I'll bring that up again.

We did get the Wolfhounds out. I remember the CO of the Infantry company asking which track I was in. So, I told him to look for a track where the back door was opening. He was able to make it and jump in. By this time, most of the infantry was able to pull back behind our tracks. Saber 6 (LTC Otis) told me to move Charlie Troop back from the clearing where most of the action took place. I still had an unmanned tank out there. LTC Otis said he would have B Troop sweep in from another direction and get the tank. I don't think that ever happened, but I know what I heard. It's not a good idea to lose a tank.

By now, it was getting late in the day, and we moved back to a more secure area on the other side of Highway 1. We set up a perimeter and took a well-deserved break. The break didn't last long as we went back into Hoc Mon the next day, 7 February 1968. We went back to where we last had possession of our tank. It was still in the clearing. We discovered that the enemy had booby trapped the tank with explosives. That was pretty much the end of our use of that tank. We came under heavy fire again from the enemy and unfortunately lost four more men.

We spent the next couple of days attached to the 2/27th Infantry. My troop was split up and attached to three of their companies. I flew around in a chopper and just advised when I could. Luckily, we did not encounter any of the enemy.

I was able to get control of the troop on 10 February. We had to return to Hoc Mon, but this time we came into the clearing from the north. Once again, we came under heavy fire. First platoon took the only casualty. Really sad, as I knew SGT Kenny Devor. Damn fine trooper. I had asked his best friend in the unit, Richard Travato, to be my driver on the C6 track. Richard and Kenny had been together through basic and advanced training. I will never forget his demeanor when he heard SSG Kinnard radio me that Kenny was hit. It was very sad.

To finish up my experience during February 1968, the troop was involved with two or three more skirmishes with the Viet Cong. There were additional casualties. I remember one day we were traveling up a very dusty road. Visibility was the pits, and we couldn't move very fast. LTC Otis asked me what the holdup was, and I told him we couldn't see where we were going. Shortly after that we came under enemy fire, small arms and rockets. One tank got hit and we lost C26, SFC William Jenkins. He had only been with the troop for a short time. Between 12 February and 21 February 1968, we lost five troopers.

These were bad times. We were out on maneuvers most of the time and rarely caught a break. I was doing my best with little sleep and really relying on Glenn Otis to show me the way. He was gone one day, and I made the worst decision while I was CO. It was getting late, and I was told to bring the troop back to the squadron field camp.

On the map, the shortest route was through a small, wooded

area. I decided to take that route, a mistake. We were just barely into the woods when a rocket hit the lead tank. It was disabled and we came under heavy fire. We were able to break contact, but the tank was lost. That makes tank #2 that was lost while I was Troop Commander. And the driver was lost as well. I will always believe that had Glenn Otis been in the sky that day, I would have been advised to take a longer but safer route back to our field camp.

I want to say a few words about LTC Glenn Otis. I think all troopers ever assigned to the 3/4 Cavalry will agree that Glenn Otis was the most beloved of all the Squadron Commanders. Even now, when we hold our reunions, the only commander whose picture is displayed is the one of Glenn Otis. He was a LTC when he commanded the squadron, but when he retired from the Army, he had four stars. He was one fine gentleman and a great leader. He always thought of the troopers and their welfare. I'm so glad I got to see him at our Nashville reunion.

The end of my tour was coming up and I was scheduled for R&R in Australia. LTC Otis chose CPT William Shaffer to take over C Troop from me. Bill was a good guy. I will always consider my time with C Troop and the squadron as my favorite time in the Army. It was exciting, dangerous, and scary, but the troopers I served with were an awesome group of men. We lost far too many and I sometimes wonder if it was worth it. Watching the *Vietnam War* on PBS makes me think it was all a huge mistake.

STEPHEN L. PRESSER

BRANCH OF SERVICE: U.S. Army, Combat Medic
DATES OF MILITARY SERVICE (ACTIVE AND RESERVES): September 1967 to June 1970
UNITS SERVED WITH IN VIETNAM: Headquarters Company, 2nd Battalion of the 16th Infantry, 1st Infantry Division
DATES OF SERVICE IN VIETNAM: March 1968 to March 1969
HIGHEST RANK HELD: SP5
PLACE AND YEAR OF BIRTH: Blytheville, Arkansas – 1944

AMBUSH BATTALION

My battalion was known as the ambush battalion for much of my tour in Vietnam. On one occasion, my company set up an ambush surrounding a small village because of suspected NVA in the area. My platoon set up behind a berm in the darkness for cover.

After a while, we spotted movement through the night vision goggles. We then started shooting up flares and there the NVA were, no more than 50-75 yards away, heading straight toward us. All hell broke loose. We started using all our fire power, to include machine guns, RPGs, and our M-16s. The NVA did the same.

Being a medic, I was not required to fire. However, I did use my M-16 since nobody was hollering medic. It was very chaotic, and we were beginning to use up our ammunition and had run out of flares. Nobody ran and we stood our ground, even though we thought we would not survive.

Our soldiers were shouting back and forth about the position of the NVA as to where we should direct our fire, and nobody panicked. Evidently, the other platoons were told to hold on to their positions, so there was no relief. If not for the protection of the berm, then it would have been worse, but when we thought it was all over for us, our APCs took up position behind us and fired their 50 caliber machine guns.

They saved our lives, because the NVA turned and ran. We made a sweep of the area afterwords and the NVA did leave some of their dead and wounded behind.

Amazingly, we had no casualties! We were all very thankful that we had survived and shouted out our survival very loudly! It was then that I realized the bravery of the American soldier. I look back on this battle now and it makes me proud to be an American. It was this kind of grit throughout our history that has made this the great country we have today! Freedom is not free!

BARBARA BECK PRYOR (JAY'S WIFE)

JUST MARRIED

I stepped off the plane, the sun-warmed breeze touching my face. I was home, that is, it was my first place to set up housekeeping with my new husband, Jay. We had married in Atlanta, honeymooned in Gatlinburg, and then had flown Delta to Hawaii. What a strange mixture of feelings: being a newlywed, landing in a paradise to live so far from what I had known as home for all my life, and then the immensity of knowing my gentle husband and so many other young men and women were fighting a war. The Vietnam War.

We were greeted by one of Jay's fellow officers, Roger Strait, who put gorgeous, fragrant leis of plumeria blossoms around our necks. It was evident from the beginning what strong ties these men were building. Their friendships would last forever. Each of them depended on the others with the jobs they did aboard the USS Hissem (DER-400), a small but formidable ship which was home ported in Pearl Harbor.

What led us to this new life? Jay had decided that, as he approached graduation from UGA, he would apply for Naval Officer Candidate School in Newport, Rhode Island. Everything was all business there, but he was not in a war zone yet. I went to OCS graduation with his parents, Wilda and Gatewood. How very proud they were of their son and his accomplishments. In our guest room, a framed photo shows Jay and his dad in front of the school and the importance of that day for both men.

That was the first step that led us to Hawaii. Then there was Christmas eve, when Jay was home on leave. We went to Emile's French Café in downtown Atlanta. Jay proposed! I did not see that coming. Neither did my parents. It seems that when Jay arrived at my home that evening and was sitting in the living room, he said, "May I ask Barbara to marry me tonight?" My mom shouted out "Tonight? You want to marry her tonight??"

Moving on, the first thing we had to do in Hawaii was find a place to call home. While Jay was aboard the Hissem, I would hit the streets (in our car borrowed from Roger) before the sun was up, get the morning paper, and circle any possibilities in the classifieds. Then I arranged to see all of them that day if possible.

Apartments were in demand. I saw a number of sketchy places, even one where a man was sleeping. Then, of all places, one turned up in the building across the parking lot from where we were being put up temporarily by the Navy. I called the landlord, Mr. Walters. Even though I was risking losing this apartment, I told Mr. Walters he must provide a double bed, a small table, and two chairs. He said yes! Perfect.

I loved it as our first place to set up housekeeping, but it had its challenges. The first night, Jay happened to be on ship, so I was there putting things in the linen closet and our closet. To admire my work, I opened the linen closet, and roaches were running all over the shelves! From then on, I kept all clothing and linens in suitcases.

I do believe that certain things show up on my path that were just meant to be. That happened with the neighbors, Elaine and John Kross, who moved in next door right after us. John was also assigned to the Hissem as an Ensign. We became the very best of forever friends.

One of the most memorable times we shared was in their apartment watching the first landing on the moon. Then there was the time that John turned on the oven to preheat, not knowing that was where Elaine stored items like bread (seemed like a smart place due to the roaches).

Remember the two chairs from Mr. Walters? When Elaine and John came over to our place, they had to bring their two chairs. One day the Hissem was on maneuvers somewhere off the south shore of Oahu, so Elaine and I drove to an area where there was a beach, and we could see a ship off in the distance. We asked the lifeguard if he could make out with his binoculars what ship it was. "An even 400," he shouted. Yes, that was the Hissem.

The Hissem was just one mode of transportation for Jay; we had given Roger his huge sedan back after buying a used Corvair with dual carburetors. I can remember there were two carburetors because I took the car to the Navy garage on base to have them instruct and help me with oil changes and repairs. It was wonderful getting to do a little bit of that work myself and save some money.

Even with the ship's demanding schedule, we managed to leave Oahu and fly to two other islands (Kauai and Maui), climb barefoot to the chest of Sleeping Giant on Kauai, drive the infamous Hana Highway curving in and out high above the Pacific, investigate the entire coast of Oahu, have nights out for suppers and musical performances, spend time at Hanama Bay and Sunset Beach, have supper on the Hissem when Jay had the duty, and attend both formal evenings and casual cookouts with the other officers and their families.

Jay and I had married on July 6. The following November, just before his ship left for the Western Pacific, we shared our first Thanksgiving. And also, it was the first time I had stuffed and baked a turkey. All 16 pounds of it. That was the smallest the commissary had.

The Hissem legacy lives on today as her men gather periodically in different cities across the United States. The stories they share are sometimes heartbreaking, sometimes funny. The men are cemented together in a brotherhood that will never be broken.

My memories of the Hissem and her men are stored safely in my thoughts, although the Hissem was "retired" at sea and the last Commanding Officer Jay served under on that ship, Rear Admiral Morton E. Toole, has passed on and was buried at Arlington Cemetery on December 9, 2025.

It is my privilege and joy to be part of the Navy family.

THOMAS A. (TOM) ROSS

BRANCH OF MILITARY AND JOB HELD: US Army Special
Forces — Operations & Intelligence
DATES OF MILITARY SERVICE: 1966 — 1992
UNIT(S) SERVED WITH IN VIETNAM: Detachment A-502, 5th Special
Forces Group
DATES SERVED IN VIETNAM: January 1968 — December 1968
HIGHEST RANK HELD: Major (O-4)
PLACE AND YEAR OF BIRTH: Huntington, WV — 1945

YOU NEVER KNOW

For those who served there, the Vietnam War could provide moments of terrifying chaos or hours, even days, of utter boredom. Those who carried weapons and fought the battles knew that they were contributing to the effort to provide freedom for the people of South Vietnam. Wrongly considered, there are those who served in non-combat roles, or in rear areas, who felt then (and still feel) that because they weren't actually fighting, they weren't doing anything meaningful. Nothing could be further from the truth.

The truth is that without its support personnel and units, it is unlikely that any military force would be successful for very long without them. So, the purpose of this chapter is to encourage every Vietnam veteran, regardless of your job title, to tell your story of the time you spent in that faraway place. You would be writing not only for yourself, but also for your family, descendants, and future generations. You never know who might read what you have written. To prove my point, I offer the following story.

The story begins just days short of the deadline to submit stories for inclusion in the book you are now reading. The publisher, Bob Babcock, was pleading for members of the AVVBA to write and submit a story. During monthly meetings and in emails, Bob explained to members the importance of providing their individual pieces of the Vietnam puzzle that will become an important part of our nation's history. It was the day after one of Bob's emails when something remarkable happened.

After seeing the red, white and blue U.S. Postal truck pass our house, I walked out to the mailbox to collect whatever had been left. There were several pieces of mail and among them there was an envelope addressed to me that had a computer-generated address. When I got into the house, believing it to be junk mail, I threw it on the kitchen counter to throw away.

Well, a short while later, my wife, Amy, came to me, handed me the letter, and said that I might want to read it. She had opened it before tossing it into the trash.

When I started to read the four-page letter, I was surprised to find that it wasn't junk mail at all. Rather, the letter was from a former British SAS officer who identified himself as Wesley Richards. He said that he was writing from his home in New Zealand.

The letter was very nice and polite. Wesley explained that he had read one of my books and that he was writing in an effort to gather information about one of my Special Forces teammates. The teammate that he was researching was Sergeant Gordon Gilmore, who was a fellow British countryman who had joined the U.S. Army to fight in South Vietnam.

The letter was long, but very interesting. Wesley had many questions that I am attempting to answer for him. I did point out to him that it has been about 57 years since Gordon and I were together. Despite the time, I do remember him well. Gordon was an intelligent person with a very interesting story of his own. He served with other armies.

Wesley and I have now exchanged multiple emails and, because of something I wrote, I now have a new friend on the other side of the world.

Related to the publishing of this book, I thought the arrival of the Wesley's letter was very timely—but that was just the beginning of this story!

After reading the letter from New Zealand, I went to my computer to respond. When I opened my Gmail, there was a flood of mail, much of it was junk. Paper or electronic, I totally disregard the junk. But, mixed in with this collection of junk, there was an email from a woman who introduced herself as Chi Aga. Chi had also read one of the books I had written about my time in Vietnam. Because it named the village where she had lived, she decided she would try to find me.

As it turns out, while I was serving as an advisor at Special Forces Detachment A-502, Chi lived just a little over 100 yards from our compound. She is now in her 70's, but she told me that she still remembers how our team protected her village. In 1968, the year I was there, Chi was nine years old.

In her note, Chi said that she wanted to make the effort to find me to thank me and the other Americans who had kept her family safe. She told me that she and her family had made it to the United States safely, and that she is now an engineer and works in the Flight Controls—High Lift Systems Division at Boeing in Washington state.

I immediately responded to Chi's note. When she answered my response to her, she asked if she could call me "Uncle Tom." She said that, to Vietnamese, the title of "Uncle" was a demonstration of respect for a close senior male. Since receiving her first note, Chi and I have exchanged several emails.

Occasionally, something happens to remind me why I volunteered for service in the Republic of South Vietnam. My encounter with Chi is one of those occasions. I have been thanked for my service in Vietnam by Americans hundreds of times. But the following note included in an email from Chi, my "niece," makes it clear that at least one Vietnamese family recognizes our sacrifices and appreciates our service.

Good afternoon Uncle Tom, How have you been? I hope you and your family are doing great.

This note is to let you know that we, the Vietnamese people from Dien Khanh, Khanh Hoa, Vietnam, want to extend our heartfelt gratitude for your service and sacrifice during the Vietnam War. While our histories are intertwined with complex emotions, we recognize the sacrifices made by all who served, including you, Uncle Tom.

"Your bravery and commitment during a challenging time have left a lasting impact. We acknowledge the courage it took to serve in a foreign land and the challenges you faced both during and after the war.

"Thank you for your dedication and for being a part of our shared history.

With respect and appreciation,
Chi Aga and Her Entire Family

After reading what Chi had written, I am compelled to say—that note touched my heart. Both communications just described arrived on the same day. Amazing to me!

Write—you never know who or when someone might read what you have written and what it might mean to them.

ALAN SMITH

BRANCH OF SERVICE: US Army Infantry/Aviator
DATES OF MILITARY SERVICE: October 1965 through April 1986, all active
UNITS SERVED WITH IN VIETNAM: 74th Recon Airplane Company (RAC) and the Command Airplane Company (CAC)
DATES SERVED IN VIETNAM: Feb '68 – Jan '69 and Jan '71 – Jan '72
HIGHEST RANK HELD: Lieutenant Colonel (O-5)
PLACE AND DATE OF BIRTH: Honolulu, Hawaii – 1947

157 FLIGHT HOURS IN 23 DAYS

As a surfer kid in Southern California, I always wanted to fly airplanes. My father was a Marine Corps Major who had flown in the Pacific in WWII and in Korea during that conflict. Unfortunately, he was killed in a training accident in 1954 while making the transition to Jet fighters.

In 1965, after a very short time as a math major at Orange Coast College in Costa Mesa, CA, I joined six of my friends and enlisted in the Army at the ripe age of 18. After Basic Training, at Ft. Polk, Louisiana (garden spot of the world), I went to Ft. Benning for Infantry Officer Candidate School, graduating in December of 1965 as a shiny new 19-year-old Second Lieutenant.

After a brief stint at Ft. Ord, California, I went to Officer Fixed Wing Aviator training at Ft. Stewart, GA, graduating in December 1967. Because I did fairly well in flight school, and all the Mohawk training slots were filled, advanced aircraft training was in the U-8D, Twin Bonanza. At that time, we were all in flight school on orders to Vietnam, with Temporary Duty to flight school.

I arrived in Vietnam in mid-February 1968. The TET offensive was still going on. My orders to Vietnam were to the USARV Flight Detachment to fly the U-8 aircraft. When I got to the battalion, they informed me that the unit was now the Command Airplane Company (CAC) and that the U-8s were gone, replaced by the brand-new U-21s, an unpressurized version of the Beechcraft King Air twin turboprop.

With that, they sent me to Phu Loi in the III Corps area, to the 74th Recon Airplane Company (RAC), flying the O-1a Birddog. A small single engine, two seat Cessna airplane, built for the Army during the Korean War. When I got to the unit, they assigned me to a room. My roommates were Nolan B. (Mac) MacMorris, also known as The Black Baron, and Bill Donaldson whom I had met during Flight School. Mac told me that they had been being attacked almost every night with mortars and rockets. He said if that happens, I should get under my bunk, and he'll tell me what to do.

Well, it did happen, and when I tried to get under my bed, that didn't work as there was a footlocker and other stuff under the bed. So, I managed to get between the springs and the mattress. The next thing I saw was a pair of white jockey shorts running out the door. I then heard a shouted, "Un-Ass this place man!" and he was gone.

I threw on a flack vest and a steel helmet, grabbed my M-16, and followed the crowd to a large bunker, known as the White Elephant. In the chaos in the bunker was big guy hollering, "If you have a helmet, a flack vest, and a rifle, get over by the door for a reaction force. They're expecting a ground attack."

I had no idea who that guy was, but he acted like he was in charge (he was The First Sergeant). Like a good soldier, I ended up being the first guy by the door. A few minutes later, the Commander came in and saw me, he immediately said, "Aren't you Smith, my new pilot?"

"Yes Sir," I responded. He said, "Get out of the door, that's not for the pilots." So went my first night in Vietnam.

I was given the Call Sign ALOFT 47. The program in the 74th was that you would fly three day-missions, then three night-missions, and continue until you had 140 hours. Then you had three days off before doing it again. I hit the magic number in twenty days.

On my first day off, I was awakened and told that the Commander wanted to see me ASAP. I reported to the CO, and he immediately said, "Who do you know? Is your daddy a Senator or something?" I told him that my father had long passed and that I don't know anyone. He said, "You must know somebody." I finally said, "Sir, what are you talking about?"

He then said, "In three days you're going over to the Command Airplane Company at Long Thahn North and they're going to check you out in the new U-21." I asked if I had anything to say about it and he said, "No!"

The 74th RAC had air-conditioned rooms, the CAC did not. He said, "That's ok, just sleep with your flight records, you won't stink." Then he said, "And by the way, screw your day off, you're flying tonight." Hence, another 17 hours over the next three days.

What I didn't know at the time, guys were extending their Vietnam tours to try to get into the U-21.

In a moment of insanity, a few weeks after learning to fly the U-21, I asked the battalion commander to send me back to the birddog unit. He looked hard at me and said, "Smith, if you ever say that again, I'll have your head examined."

I spent the rest of my tour flying VIPs of all sorts all over Vietnam. We routinely flew in and out of 63 different airfields throughout the country. The Major at the 74th was partially correct, we did have air conditioning, but my flight records looked pretty good, about 1,000 hours of multi-engine turboprop time.

CARL H. "SKIP" BELL, III

BRANCH OF SERVICE: U.S. Army—Armored Cavalry Officer, Aviator
DATES OF MILITARY SERVICE: 1967 to 1998
UNITS SERVED WITH IN VIETNAM: A Troop, HHT, B Troop 1st
Squadron, 4th Cav, 1st Inf Div (First Tour); C Troop, 3rd Squadron, 17th Air
Cav; 18th Corps Aviation Company; G3, HQ 1st Aviation BDE (Second Tour)
DATES SERVED IN VIETNAM: Feb 1969 to Feb 1970; Feb 1972 to Feb
1973
HIGHEST RANK HELD: Colonel (O-6)
PLACE AND DATE OF BIRTH: Decatur, GA—1945

ANOTHER FU LIZARD STORY

In May 1969, I was a Platoon Leader in A Troop, 1st Squadron, 4th Cavalry, 1st Infantry Division (the Big Red One). We were part of a task force operating under the control of the 1st Battalion, 16th Infantry. The task force consisted of the Mechanized Infantry Battalion, the Armored Cavalry Troop, a 105 mm Field Artillery Battery, and an Engineer Rome Plow Company.

Our mission was to open a road that had not been used since the French left Vietnam in the 1950's. Jungle had completely overgrown the road (hence, the need for the Engineer Rome Plow Company). The road was in the foothills of the Central Highlands, and it ran northeast along a ridgeline from the Montagnard village of Dong Xoai (north of the Phuoc Vinh basecamp) to an old rubber plantation called Bunard (which was also an American Special Forces camp) and then back to the west to the Song Be basecamp.

The Special Forces Camp at Bunard could only be resupplied by air, which was one of the reasons for clearing the road. The area was mostly jungle and was thought to be inhabited by numerous NVA units.

Several things of note (to me) occurred during that operation. First, I traded a Montagnard tribesman in Dong Xoai some C-Rations for a net hammock which I strung up in my Armored Cavalry Assault Vehicle (called an ACAV—an M-113 armored personnel carrier that had a .50 caliber heavy machine gun and two M-60 machineguns mounted in gun shields on the top of the vehicle and sandbags in the bottom—we used it like a light tank and fought from the vehicles whenever possible, though when we were in bunker complexes or operating in ground or streambeds too soft for the vehicles, we became Infantry and fought dismounted). I strung the hammock up next to my radios between the driver's compartment and the back ramp hook of the vehicle and I slept there every night we were in the field (which was most nights) or not out on ambush between that time and the end of my first Vietnam tour.

Another interesting thing was that I found I could sleep through outgoing artillery in that hammock, even though the differential pressure from the guns firing over my vehicle would sometimes literally lift me up in the hammock.

We provided perimeter security to the Field Artillery Battery and Engineers in the fire support base/Engineer Rome Plow base that we cleared out on one of the hilltops (more on that later), and when the Artillery Battery received a fire mission, they would fire over the tops of our tracked vehicles.

Outgoing fire going over my vehicle would literally lift me and the hammock up and then drop us back down the way we were. I would be so tired from patrolling all day that I could sleep through that jostling caused by the outgoing artillery, but I would wake up when I heard my callsign on the radio speaker (which was right next to my head). Human beings have a wonderful quality to adapt to things.

A third interesting thing was that it takes 900 sandbags to

make a U-shaped revetment for an M-48A3 tank and a lesser number to make a revetment for an ACAV. We had both tanks and ACAV's in my platoon (three tanks and seven ACAV's when all the vehicles were up and running).

The Infantry Battalion Commander decreed that all the positions for the vehicles had to be sandbagged in (which was a hell of a lot of work). We tried to tell the guy that one of the ways to decrease the likelihood of an attack on our position was to move the vehicles to different places on the perimeter every night to throw off any reconnaissance the enemy may have done in preparation for an attack (the NVA were very good and very methodical most of the time when it came to attacking a fire support base — they liked to know where all the heavy weapons, machineguns, etc., were in advance, and when those weapons (or armored vehicles) were moved around it confused them).

He was not impressed by our arguments, so we ended up with 'permanent' positions for our vehicles (plus a lot of sore backs and blisters from digging up that laterite ground to fill the sandbags). The bad guys also knew where every armored vehicle would be parked in our fire support base every night. War is hell.

The fourth interesting thing (and the reason for this story) was that I discovered the existence of a species of lizard that I had never heard of. The troops called this little fellow a "F*ck You Lizard." We ran into all kinds of creatures (some lethal, some painful, some funny) in the jungle, including cobras, bamboo vipers, swarms of stinging ants, large scorpions and centipedes, tigers and monkeys.

The F*ck You Lizard was harmless, but it could be pretty scary the first time you ran across him (or her, not sure which). My first experience with this lizard happened on the operation I'm writing about. As was mentioned previously, there were a LOT of NVA operating in this area, and they had not been messed with in some time, so they were lax about leaving things like green plastic ponchos and lit cigarettes on the ground when we approached where they were.

Of course, when we found those things, we got very nervous. I mentioned previously that the road we were clearing had been closed since the 1950's and the jungle had literally grown over the road. As we moved up it securing the Engineer Rome Plows as they knocked down the jungle, we came upon a French army truck full of bullet holes with a skeleton in the driver's seat wearing a steel helmet.

The concept of the operation was to clear enough of the road to get through to the hilltop that was to be used as an artillery fire support base/Engineer Rome Plow base area and then work back up and down the road clearing several hundred meters on both sides so that the road could be used as a main supply route between Phuoc Vinh and Song Be.

The first night we were on the hilltop that eventually became the fires support base/Engineer base, we had a fairly small perimeter. We had knocked down the jungle that had been on that hilltop and some of that jungle included some large bamboo that we also knocked down. We set up the perimeter and my Troop Commander (a great guy named Bill Newell) ordered that if we did any harassing and interdiction fires (random shooting to discourage the enemy from probing our perimeter), that those fires not be done by our machineguns or M-16's, but by M-79 grenade launchers.

The grenade launcher had less of a signature muzzle flash and so would not give away our positions to the enemy (note that this was BEFORE we sandbagged the vehicles—this was our first night on the position).

After it got dark (and it got seriously dark that night—there was no moon and low cloud cover), the temperature also rapidly cooled down and this caused the bamboo that we had knocked down earlier in the day to contract and make snapping-popping noises which sounded to me like people moving around the perimeter. On top of that, suddenly we heard something that sounded like, "EEE EEE EEE EEE EEE EEE AAH AAH AAH AAH AAH F*ck You F*ck You F*ck You!" coming out of the jungle. I had visions of World War II movies about the Marines fighting the Japanese in the Pacific and the Japanese soldiers yelling "Marine you die!" and stuff like that.

It turns out that the noise was being made by a lizard, but none of us knew that at the time. The bamboo cracked and the lizard(s) yelled well into the night. I had a case of M-79 HE (High Explosive) grenades on my vehicle and I (and several others around the perimeter) began firing the grenade launcher at the sounds. I went through that whole case of M-79 grenades that night and my shoulder was black and blue.

We would pop the occasional hand flare and (of course) we saw nothing, but we also knew that NVA sappers were very good at getting into defensive positions at night, so we did not get much sleep. We found out later from the Special Forces guys up at Bunard that it was a lizard that made that F*ck You call.

One of the things I used to joke about was wanting to take one of those lizards home from Vietnam with me as a pet. The other thing I wanted to take back was a rubber tree, so I'd have a place to go tee-tee (we spent a lot of time in rubber plantations).

Combat can be a terrible experience, but it also has its humorous side (and this is one of them). The other aspect of combat is that it binds people together like no other experience that I'm aware of. It is what brings Veterans together in organizations like the Atlanta Vietnam Veterans Business Organization (AVVBA).

Our 'lowest common denominator' is that we all served in that war, even though we had very different experiences and lifestyles while we were there—that service in Vietnam is what binds us together and what makes the AVVBA the very special group of people that it is.

NORMAN E. ZOLLER

BRANCH OF SERVICE: U.S. Army: Field Artillery Officer (1962-1969); Judge Advocate General Corps Officer (1978-1993)
DATES OF MILITARY SERVICE: 1962-1969 (Active); 1978-1993 (National Guard then Reserves)
UNITS SERVED IN VIETNAM: 1st Tour: Detachment B-130, Special Forces (1964-65); : 2nd Tour: HQ 3rd Brigade, 82nd Airborne Division (1968-69).
DATES IN VIETNAM: September 1964 to March 1965; March 1968 to February 1969
HIGHEST RANK HELD: Lt. Colonel (O-5)
PLACE AND DATE OF BIRTH: Cincinnati, OH — 1940

REFLECTIONS ON VIETNAM: TRAGEDY, INSPIRATION, SERVICE, LEGACY

I do not intend to comment or dwell on the purpose or the morality of the Vietnam War. Undoubtedly, barrels of ink have been spilled, and billions of words have been written about it which, for the United States, formally extended from November 1, 1955, until April 30, 1975: thus, nineteen years, five months, and twenty-nine days. But it is worth noting that the United States did not formally declare war on North Vietnam, but rather, engaged in the conflict under authority of Congressional actions such as the Gulf of Tonkin Resolution or Southeast Asia Resolution, Pub. L. 88-408, 78 Stat. 384 on August 10, 1964.

Moreover, the French had been historically involved in Vietnam and Indochina under colonial rule for a century, culminating on May 7, 1954, when Viet Minh forces overran Dien Bien Phu, effectively ending the Indochina War (1946-1954) and the resulting gap of authority which ensued thereafter.

On a personal level, however, members of the Atlanta Vietnam Veterans Business Association (AVVBA) have now shared more than four hundred stories in the *I'm Ready to Talk* series compiled by Bob Babcock concerning their first-hand experiences on the ground, on the seas, and in the air... largely about what they saw and did while deployed. In any number of respects, those stories are emotional, honest, heartbreaking, and heartwarming.

But: what of it all? What do our compatriots conclude and think about such experiences? And many of our colleagues still bear the burdens of actual physical, mental, and even moral vestiges of what occurred to them and their families, personally.

Their stories, at least partially, recount such feelings.

During the Vietnam War, sadly and demographically, there were reportedly 3.8 million casualties, nearly half of which were the deaths of civilians caught up in and imperiled in the fighting (estimated at 190,000—430,000). Tragically, of the deaths, 58,281 were United States service members (47,434 killed from combat; 303,644 wounded). In my unit alone, the 3rd Brigade, 82nd Airborne Division, two hundred twenty-seven died and more than eleven hundred were wounded over the course of twenty-two months of combat.

Of those while I was there for a year beginning in March 1968, more than a hundred of my fellow soldiers were killed during the first three months. Enemy losses that year (February 1968—February 1969) included one hundred twelve Viet Cong (VC) prisoners of war, two hundred sixty-seven VC killed in action, and seven hundred seventy-six North Vietnam combatants killed in action.

Further, in some places, significant protest about this conflict occurred as these were tragic, troubling times abroad and at home.

In Vietnam, we Soldiers, Sailors, Marines, Air Force, and Coast Guardsmen and women discharged our responsibilities, protected, and fought side-by-side, whether in the jungles, villages, hamlets, supply centers, headquarters units on the ground; in the air; or on the rivers and seas. For those gallant and heroic efforts in my Army Airborne Brigade alone, there were warranted recognitions with two awards of the Distinguished Service Cross, five awards of the Distinguished Flying Cross, fifty-seven Silver Star Medals, one hundred ninety-five Bronze Star Medals with "V" (Valor) device, and twenty-five Soldier's Medals. And in the accompanying citations of these awards, there was narrative commending and attesting to their inspirational deeds of gallantry.

These soldiers did not seek personal recognition; they simply saw their duty and did it. And our unit and its soldiers were not unique in these respects to the commitment they perceived to their fellow comrades. Over the years, more than 9 million United States men and women sacrificed and served during the Vietnam War era. This includes those who served within the borders of South Vietnam, those who served in Southeast Asia, and those who served elsewhere in direct or indirect support of that engagement. *[Editor's note: According to the website "Statistics – Vietnam Veteran Project," during the Vietnam War, approximately 9,087,000 military personnel served on active duty from August 5, 1964, to May 7, 1975. Among them, about 2,709,918 Americans served in Vietnam.]*

It was they who continued and perpetuated the tradition of service, sacrifice, and inspiration to the military, to the Nation, and to its future generations.

So, what are the notable bequests, some of which are of an enduring nature, particularly from our AVVBA colleagues, who served in Vietnam and Southeast Asia?

One *legacy* is the increasingly renowned AVVBA-produced documentary video, *Truths and Myths about the Vietnam War*, which at this writing has been viewed by more than a million people, not only in this country, but also from abroad. Moreover, it has been or is scheduled to be shown throughout the nation, including on one hundred-eighty public and commercial broadcasting stations, six of which are in Georgia.

Another legacy is the *service* that has been and continues to be via the recurring Project Mail Call where AVVBA members, and other citizens, gather at the home of Ed and Mary Ettel to pack and send care packages with hand-written messages to active-duty service members, principally to those stationed abroad. These have been gratefully received and often acknowledged with return appreciation messages from the military recipients.

Another is the *service* by many AVVBA colleagues who meet, greet, and provide personal support and encouragement to active-duty service members while at the USO Reception Center at the Hartsfield-Jackson Atlanta International Airport. On any given day, there can be hundreds of such service members from all military branches who are passing through Atlanta en route to domestic training sites or deployment to foreign duty stations.

Finally, and perhaps the most *enduring legacy* are the educational scholarship programs through the AVVBA Foundation to which thousands of dollars have been generously and charitably gifted by AVVBA members to Georgia universities and institutions of technical or higher learning. By such gifts, the next generations and inheritors of the AVVBA, will benefit from our experiences and will share and pass on their experiences and own legacies to *preserve, protect, and defend the Constitution of the United States* and this *honorable and noble Nation against all enemies foreign and domestic.*

Decades ago, we first took and pledged our respective oaths of commitment, we now recall those sacred words with and for our successors: *so help us God.*

WILLIAM E. TURNIPSEED

DATES OF MILITARY SERVICE: September 1968 – April 1971
UNITS SERVED WITH IN VIETNAM: MACV, IV Corps, Vinh Long Province
HIGHEST RANK HELD: First Lieutenant (O-2)
PLACE AND YEAR OF BIRTH: District of Columbia – 1946

A RICE SEASON

I graduated from the University of Virginia in May 1968 with a degree in International Relations. My plan was to attend George-town University and obtain a master's degree and join the Foreign Service. Those plans went awry when it became apparent that the Vietnam War was not going to be over by the date of graduation and I would be subject to the draft, having then lost my deferment.

I am the son of a career Naval Officer who joined the Navy as a "white hat" in the heat of the depression, was captured on Corregidor, spent three years as a POW in the Philippines, mostly in Cabanatuan prison camp north of Manilla (made famous in Hampton Sides' book, "Ghost Soldiers"), and commissioned as an Ensign out of the War. He retired as a Commander in the Navy in 1958. He was in communications.

I attempted to get into Navy ROTC (the two-year program) in the spring of 1968 before graduation but failed the physical due to "poor color perception." So, I enlisted in the Army for OCS after graduation. The Army had no color perception requirement. In fact, it was never mentioned in my draft physical. I practically froze to death at Fort Dix in the winter of 1968-69 in Basic Training and AIT; and I practically died of heat exhaustion at Fort Benning in OCS before being commissioned as a Second Lieutenant in the Infantry on August 9, 1969.

My first duty station was Fort Carson, Colorado. Worked as an Assistant S-4 at Brigade, and then a platoon leader before receiving orders to report to Vietnam on April 4, 1969, flying from Travis AFB in Oakland, CA, to Hawaii, Guam, Clark Field in the Philippines, and Ton Son Nhut, South Vietnam. I was assigned to MACV, attended a three-week Vietnam school in Saigon, before reporting to Vinh Long, IV Corps, in the Delta. I was the third in my OCS class to report to Vietnam.

VINH LONG

A river town. It is located on the north leg of the Mekong River, due south of Saigon. The province of Vinh Long, in the approximate shape of a trapezoid, was bordered on the south by the southern leg of the Mekong River. Across the river was Can Tho, the largest city in the Delta, and IV Corps Headquarters. Both legs of the Mekong eventually emptied into the South China Sea.

Vinh Long province was the site of many engagements with both VC and possibly regular North Vietnam cadre by the 9th Infantry Division and ARVN in 1967 and 1968.

I was initially assigned to a MAT Team ("Mobile Advisory Team") in a village located about five or more miles from Vinh Long City, which had the headquarters of MACV in the province, and which also hosted a PCP runway US Army airfield with Bell Hueys, gunships, and single and multi-engine fixed wing aircraft.

A MAT Team's military mission was to work with the local village or county chief to devise defensive schemes to incursions by the Viet Cong, and to train "rice farmers" how to employ those schemes for their own protection and preservation of their crops.

Composed of two officers, two NCO's, one of which was a medic, and an interpreter, the MAT Team was the embodiment of the belated plan to "win the hearts and minds" of the populace in the countrysides in order to nullify the Viet Cong who preyed on rural villagers, extracting taxes and food to survive and who carried out military missions against the South Vietnamese government.

This was the latest plan of the US government after the election of President Nixon to "Vietnamize" the war, and hopefully "Vietnamize" the peace as well. Tet had concluded in 1968 as a military failure of the VC, but a propaganda victory for them. Political pressure was being exerted on President Nixon and Henry Kissinger to lower American participation while at the same time enhancing both uniformed and civilian "soldiers" in their fight against the VC in South Vietnam.

Vietnamization was led by retired Colonel John Paul Vann who had spent multiple tours in Vietnam and who led the charge to win "the hearts and minds."

When I arrived in Vinh Long, the rice fields were being planted. Hundreds of farmers bending at the waist would insert plugs of rice plants into acres of paddies awash with river water introduced by locks. Water buffalo were the chosen "beasts of burden."

When I left Vietnam one year later, I had witnessed a harvest of the rice. For this area of the Delta, rice was the economic engine of commerce. And while villagers with whom I lived had plenty of "needs," food was not one of them. Rice and fish, the staples of the diet, were plentiful.

Military preparedness was however in short supply. While the District or "county" troops in Vinh Long were outfitted with both M-16's and grenade launchers, and conducted operations throughout the county, sometimes in coordination with the ARVN, the villages were lucky to have two or more "police officers" to keep the peace and to ward off the VC if in the area. Some may have been armed with M-16's, but most were outfitted with carbines, M-1's, and the occasional BAR.

A MAT Team was assigned to a village for a couple of months, long enough to develop a "defense" plan for the village in the event of an attack by the VC before moving on to a new village to do the same thing. While in any village, we would go out on "operations" with village police which would often result in "lunch" at the home of some potentate where we would end up drinking Bac Si De (a potent rice wine) in the interest of international relations, winning the "hearts and minds."

We often took villagers to the airfield and set up a shooting range in an effort to test their skills in the use of these WWII weapons, assuming we could get ammunition for those weapons. We would also test their skills at cleaning their weapons, and "command and control" in the event of contact with the VC. Candidly, these efforts were often short lived. Trying to turn rice farmers into even "amateur" soldiers was absurd.

We moved to three or four villages in the year I was in Vinh Long. Each time we built our own Team Bunker with PCP, sandbags, lumber, and other matters "stolen" from the airfield. We were lucky to have a hooch provided to us by the village to which we were assigned that included a concrete floor. We were provided a propane refrigerator and stove top. We purchased food at the airfield PX and either prepared it ourselves or had it prepared by a lady hired in the village.

Although there were "hotter" provinces in the Delta than Vinh Long, we were not bored by the lack of "action." In addition to getting to know the village officials, learning the geography of the villages, and developing village defense plans, we often accompanied village police on forays into the rice paddies to check out reports of VC incursions.

I once accompanied a District team on a night mission following twenty or more soldiers armed to the teeth into the dead of night toward a little populated part of the village. It was slow going, walking and frequently falling off narrow dikes and log bridges over streams and other water hazards.

About two or three hours into the walk, we stopped near a stream. A US captain who also went along and I squatted next to the stream bed when tracers cut through the night over our heads in a terrific racket. Our troops to our rear, hidden by the night, opened up and the fire fight was over in a matter of seconds. We later located our assailants, three now dead VC who mistook the troops to our front as their only foes. The troops to our rear finished them off in short order. We medevaced one of our troops who had suffered a minor flesh wound.

But we also had the opportunity to engage in humanitarian endeavors assisting individual villagers with medical problems, some congenital, some accidental, and a few of military origin. In Vinh Long town was a corps of US Air Force surgeons who worked day and night on military and non-military injuries. We often took in children with cleft palates for repair. I can recall we took in a teenager with an arm attached to his trunk by skin which had melted when he was scalded by hot water. The surgeons cut and discarded the loose skin impeding use of his arm.

Toward the end of my tour, the colonel in charge of our MACV presence in Vinh Long tried to convince me to extend six months since I was scheduled to ETS out of Vietnam. He offered to make me a captain upon extending. I politely declined, advising him that I needed to return to school out of the Army. I am sure he was not surprised.

In my last two weeks in Vietnam, my MAT Team having been

disbanded, I was sent to an ARVN encampment in the south part of the province. I lived in a pup tent with a couple of other American soldiers attached to an ARVN battalion calling in the occasional air strike from aircraft in support of the ARVN troops in the distance.

Saying so long to my long-standing medic, SFC Harry Faengler, and my assistant, LT Bill Pasko, I took a plane to Fort Lewis, Washington, and from there a plane to McGuire AFB next to Fort Dix where I ETS'ed out of the Army six months early.

The Army was cutting back, and since I had six-months or less to serve, I was "riffed." Ironic. I left the Army from the same post in which I had entered in Basic Training two- and one-half years earlier.

As I ate breakfast with a couple of other LTs in the mess hall that morning, in the presence of new recruits just coming in, I could not help but thank my Creator for my good fortune to have survived the year. I was proud to have served with my MAT teammates that year, and hoped that we had accomplished some good things, although mercifully spared the carnage experienced by our fellow G.I.'s in other theatres of the war.

EPILOGUE

I left Vietnam with the suspicion that my comrades and I had been the beneficiaries of Nixon's Cambodian bombing in the spring of 1969. The conventional wisdom at the time was that this bombing in the Parrot's Beak of the Delta cut down the number of arms coming into the VC from the Ho Chi Minh Trail, which in turn cut down the VC activity in Vinh Long and other provinces in the Delta.

If true, another irony is that I had always been critical of Nixon and Kissinger for failing to end the war "with honor" (remember their promise?) earlier which would have lessened the number of American casualties. But their aggressiveness in the bombing in

Cambodia may have kept me from becoming one of those casualties. They may have literally "saved my bacon."

CARL H. "SKIP" BELL, III

SOMETIMES THERE ARE NO GOOD DECISIONS

On 25 June 1969, an Armored Cavalry Troop of the 1st Infantry Division was engaged in what became a three-hour battle with an estimated battalion of North Vietnamese Army soldiers. We suffered a number of casualties in the initial minutes of the fight (we were in a night defensive position (NDP) and were attacked repeatedly with mortars, RPG's, small-arms, and machinegun fire, and several ground attacks throughout the night). The enemy broke contact at daybreak.

During the course of the battle, the friendly wounded were brought into an aid station that had been established in a ditch in the middle of the NDP. I had called for a medical evacuation helicopter early in the fight (when I knew that we had wounded), and the aircraft came on station in a very short period of time. When the aircraft arrived, the pilot told me he was willing to come into our position and pick up the wounded, but I told him not to come in yet because the ground fire was too intense, and I didn't think he would be able to come in and get out without being shot down.

At some point I crawled over to the ditch where the wounded were and yelled down to them to hang on, that a Dustoff (medical evacuation helicopter) was on station, and they would be taken out as soon as it was safe to bring the aircraft in — at that point the bullets flying around us sounded a lot like bees. As soon as I said that my Troop Medic, who was attending one of the more seriously wounded Soldiers, looked up and me and said, "It's OK, Sir. He just died."

I wonder to this day whether my decision to delay bringing in the Dustoff was the correct one — we were taking an incredible amount of incoming fire, and I felt sure that the aircraft would not have been able to get in and out with the wounded without being shot down.

On the other hand, had I brought the aircraft in when it first arrived and they were able to successfully extract the wounded, the deceased Trooper might have survived. Sometimes there are no good decisions.

DON TRAUB

BRANCH OF SERVICE: U.S. Army, Adjutant General Corps
DATES OF SERVICE: 1965 - 1992
DATES IN VIETNAM: March 1968 – March 1969, 101st Airborne Division
RANK IN VIETNAM: 2LT – 1LT
HIGHEST RANK HELD: LTC (O-5)
PLACE AND DATE OF BIRTH: Cincinnati, Ohio – 1943

THIS ISN'T WHAT I THOUGHT IT WOULD BE

So here I was, dumb and happy. It was March of 1965. I had just graduated from the University of South Florida, married to my High School sweetheart six months ago, and didn't have to take ROTC. Why would I march around the track field when the St. Petersburg beaches were only a few miles away? We just bought a house, and I landed a job with Honeywell at Tampa. Life was good!

Things abruptly changed November 24, 1965, when I received

an Order to Report for Induction, Draft Notice!! (It still hangs on the wall of my home office, known as The Bunker). Later, we found out that Florida was one of the first states to draft married men. (See the title.)

Fast forward. I finished Basic Training and, while at AIT (Advanced Individual Training), the Company Commander demanded to know why I hadn't taken ROTC. He was entirely unimpressed with my reason. He informed me they would soon be convening an OCS (Officer Candidate School) board. I was told to pick three branches, and one had to be Combat Arms. I chose Transportation, Quartermaster, and Artillery. After the laughter stopped, I was informed I would be reporting to Ft. Benning, for Infantry OCS. (See the title.)

I graduated from OCS in May 1967, and, after saying goodbye to my wife and three-month-old baby boy in March of 1968, I reported in Vietnam as an Adjutant General 2LT to be assigned to USARV HQ in Long Binh. I even had their patch sewed on my sleeve. Well after several days at the 90th Replacement Bn, I was told I was being diverted to the 101st Airborne Division, the historic division of WWII fame.

When I told them, thanks for the honor, that I was not jump qualified, they gave me their answer: No problem, the Division was being reconstituted into an Air Mobile configuration. So, I would be assigned to them as a dreaded "Leg," the pejorative term used to describe a non-jump qualified soldier. At that time, over 95% of the officers were receiving Jump Pay of $110 per month. They would continue that, even though no one jumped. (Once again, refer to the title.)

I reported to Division HQ in Bien Hoa and because of my Infantry background from OCS, I was selected to run the admin job at the forward element at Camp Eagle, about 15 kilometers southwest of Hue. The living conditions were quite austere compared to those of the Division HQ in the rear, but we actually had more privacy than most. I had my own GP (General Purpose) Small tent, the crew had GP Mediums as their living quarters, and another one for our workstation. It turned out to be a good job with a ten-man crew to provide air courier service to fire bases throughout I Corps in Hueys, LOACHs, and Chinooks — in addition to acting as liaison for any personnel issue that may arise.

My crew was proficient with basic weapon skills and easily were assimilated into our rotating turns as perimeter security. Two of our more harrowing experiences happened when we were not on perimeter duty, but when we were manning our own area of operation. We were notified that the perimeter had been breached by the VC.

The guys performed flawlessly and, according to our SOP by manning our fighting positions, consolidating all the classified documents into portable field safes and arranging thermite grenades if the documents needed to be destroyed.

Through these episodes and throughout the year, our section performed very well. We did our mission with little supervision and had only one WIA (Wounded in Action). A young private was hit in the neck and received burns from shrapnel of a mortar round that was fired at us with some frequency. Luckily, he survived. He was talking to his mother on a MARS call when it happened.

Looking back at it all, one has a strange mixture of emotions; the emptiness of missing your wife and child, relief having survived the year, hope you did everything to take care of your soldiers, and pride having been part of something bigger than yourself. I guess we all did our roles over there, and this was mine.

JEFF 'BIC' BICKERTON

BRANCH OF SERVICE AND JOB: US Army, U-1A Otter Crew Chief
DATES OF MILITARY SERVICE: 1966 - 1969
UNIT SERVED WITH IN VIETNAM: 18th Aviation Company, 223rd
Combat Aviation Battalion, 1st Aviation Brigade
DATES SERVED IN VIETNAM: 1 February 1968 - 24 April 1969
HIGHEST RANK HELD: SP5
PLACE AND DATE OF BIRTH: Queens, Long Island, New York, 1946

THE BATTLE OF DUC LAP SF CAMP

I don't know how I am writing about this. This day was one of the mind-blowing experiences of my tour in Vietnam. This happened about halfway through my time in VN in August 1968.

One morning, Monday, August 26, 1968, we flew from Nha Trang (on the coast) to a road next to Hill 722 (near the village of Dak Sak), close to the Cambodia border. There was a Duc Lap Special Forces 'A' Camp (A-239) there. It had been attacked and overrun the days and nights of August the 24, 25, and into 26.

It was up on a hill from the road we landed on. If there was a runway, I don't remember it. After we landed, a SF (Special Forces) Sergeant came up and informed us that they needed a trip to Pleiku to move the dead Montagnard CIDG (Civilian Irregular Defense Group) bodies for identification, internment, and Lord knows what. We informed the Sergeant that we could only carry 16 to 20 bodies (they had 35 bodies). It would take two trips. He said fine. I rode with him up to the camp.

Yikes, what a mess with all the holes, mortar shell casings, discarded ammunition boxes, canisters, and the smell of sulfur still in the air. The bodies of the Montagnards were all wrapped in body bags, and I helped the Sergeant and three others load them on a jeep and trailer. We took the road, drove down the hill back to the plane, and loaded them onto the floor of my 'Otter' (airplane).

Two uniformed CIDG guys came with us to accompany the bodies of their fallen comrades. All were silent, as I remember, all the way back to Hollaway Army Airfield where a SF truck was waiting for us. We fueled-up and back to Duc Lap we flew.

I had never picked up a dead man before, and I didn't start to think about it until we were on the way back with the final 'load.' As we loaded and unloaded the aircraft with these humans, I could tell rigor mortis had already set in to most of the bodies within the bags. The smell will aways be remembered.

It occurred to me that this was a serious war, a dangerous war, and it was not to be taken lightly. Those two flights were never forgotten. I never wanted this to happen to me in my remaining six months in VN. Coming home in a bag was not cool.

Note: Casualties were KIA: 6 SF, 1 LLDB (Luc Luong Dac Biet—Vietnamese Special Forces), 37 CIDG, 20 civilians; MIA: 9 CIDG; WIA: 13 SF, 7 LLDB, 80 CIDG. NVA (North Vietnamese Army) losses were 303 by body count and probably much higher. The major camp weapons at the start of the battle were two 105mm howitzers, four 81mm mortars, one 4.2 mortar, and two 106mm RRs (Recoilless Rifles).

LTJG JIM DICKSON

BRANCH OF SERVICE: U.S. Navy, Operations Intelligence Division Officer (Combat Information Center) – Officer of the Deck Underway
DATES OF MILITARY SERVICE: 1967 to 1977
UNIT SERVED WITH IN VIETNAM: USS Ticonderoga (CVA-14)
DATES IN VIETNAM: Jan 1968 to Aug 1968; Jan 1969 to Jul 1969

HIGHEST RANK HELD: Lieutenant – Inactive Reserves (O-3)
PLACE OF BIRTH AND YEAR: Lancaster, PA – 1944

YANKEE STATION – VIETNAM WAR – KOREA

I served aboard the aircraft carrier USS Ticonderoga (CVA14) for both the 1968 and 1969 deployments to Yankee Station during the Vietnam War as the Operations Intelligence Division Officer for about 60 men who worked in Combat Information Center (CIC) — the operations center on most navy ships — and Officer of the Deck, Underway.

There were normally three attack aircraft carriers on Yankee Station, called Task Force 77, that conducted air operations for twelve hours each day; one from noon to midnight, another from midnight to noon, and another one during daylight hours from 0600 to 1800 hours. This schedule provided 24-hour coverage and additional effort during daylight hours when sorties were most effective.

Our ship was operating on Yankee Station in the Tonkin Gulf when things suddenly changed on April 15, 1969, after a North Korean MIG-21 shot down a U.S. Navy EC-121 reconnaissance plane over the Sea of Japan and killed all 31 aboard. Since the EC-121 was about 90 miles from North Korea and over international waters, President Nixon ordered Task Force 71 be sent to the Sea of Japan in response — over 40 ships and the largest naval armada since WWII.

We were ordered to leave Yankee Station and become part of Task Force 71 in the Sea of Japan that included the battleship USS New Jersey (BB-62) and four aircraft carriers — USS Ticonderoga (CVA-14), USS Enterprise (CVAN-65), USS Ranger (CVA-61) and USS Hornet (CVS-12) — along with three cruisers, 22 destroyers, five submarines and six supply ships.

I will always remember reading the message traffic in the Operations Department about the response of the commanding officer of a destroyer to orders for his ship to report to Task Force 71. Incredibly, he responded that his ship would leave Yokosuka Naval Base as soon as the painting of his ship was completed—a response our Navigator Commander Duck sarcastically remarked was "from someone who has no fear of the enemy."

The size and power of Task Force 71 was intended to discourage North Korea from further adventurism and prevent this provocation from growing into a large incident. However, tensions were very high, and the Strike Operations Department of our ship—the USS Ticonderoga (CVA-14)—was ordered to draw up contingency plans to mine the harbors of North Korea with nuclear mines.

One day the ships of Task Force 71 were ordered to be closer together in formation for an aerial photograph. This operation was conducted under the direction of the flag bridge on the aircraft carriers and was conducted so poorly it easily could have caused a collision. I have never been able to obtain this photograph—perhaps, it is still classified.

After many days of high tension, Task Force 71 was disbanded and on April 25, 1969, we were ordered to return to Yankee Station off the coast of Vietnam.

ROBERT REESE

BRANCH OF MILITARY AND JOB HELD: US Army — Infantryman
DATES OF MILITARY SERVICE: 1968 to 1971
UNIT(S) SERVED WITH IN VIETNAM: B Company, 1st Battalion, 26th
Infantry, 1st Infantry Division (first tour)
DATES SERVED IN VIETNAM: Oct 1968 to Nov 1971 (two tours)
HIGHEST RANK HELD: Sergeant (E-5)
PLACE AND YEAR OF BIRTH: Atlanta, GA — 1949

RETURN TO FIREBASE RITA

As the tumultuous year of 1968 drew to a close, the Allied forces
in South Vietnam were very anxious to avoid a repeat of the surprise 1968 Tet attacks. In retrospect, it would seem unlikely that
Hanoi would tend toward any repeat as that would be, by definition, not a surprise. Nevertheless, Washington and Saigon could
not afford to take any chances of being embarrassed twice by the
same strategy.

As a result, the Allied plan was to push combat formations
further into the hinterlands to prevent NLF forces (National Lib-
eration Forces, the enemy with the Viet Cong (VC) as the military arm) from creating havoc in the larger populated cities and
towns as they did in 1968. Especially sensitive was Saigon and its
environs. A key southern terminus of the Ho Chi Minh trail fed
NVA (North Vietnamese Army) forces into northern Tay Ninh
province, only 150 miles from the outskirts of Saigon and even
closer to the key provincial capital of Tay Ninh City.

As a small cog in this master strategy, my unit — B Company,
1-26th Infantry Regiment, 1st Infantry Division was repositioned
from our regular haunts in Binh Duong province into a defensive

assignment in northern Tay Ninh, normally the responsibility of the US 25th Infantry Division.

The upper part of the province was a very underdeveloped, end-of-the-world region. The civilian populace was small, even in more peaceful times. In October of 1968, it was a wilderness occupied almost exclusively by the combat troops of the Allies and Communists.

In the last week of October, Bravo Company was airlifted into Firebase Rita, a remote outpost astride a north-south ox cart trail approximately 15 miles from the Cambodian border. Rita's sister firebases, Dot and Julie, were built five to six miles away, the three bases forming a mutually supporting triangle from which "leg" (Infantry) or mechanized units could launch reconnaissance-in-force operations into the surrounding forests.

The ultimate objective was to identify significant communist forces before they could get near the population centers in III Corps.

To complete the defense of Rita, elements of several additional units were present: Troop B of 1-4th Cavalry, 1st Infantry Division (armor); 3rd platoon of B Company, 2-34th Infantry, 1st Infantry Division (armor); Battery C of 8-6th Field Artillery (155mm); and Battery B of 1-5th Field Artillery (105mm). Aerial reconnaissance had identified two enemy base camps within 1,000 yards of Rita!

Being a newly occupied strongpoint, Rita's defenses were not yet completely built out. The perimeter bunkers were only partially finished. The concertina wire that formed the front-line defense was weaker than what you'd find at more established firebases back in our home area of operation. Ominously, the wood line had not yet been cut back to an acceptable distance, raising the danger that enemy forces could approach the base closer than was tactically advisable without being seen. It was imperative to get Rita up to par defensively as rapidly as possible.

Concurrent with the construction effort, Bravo Company 1-26th Infantry was tasked with daily reconnaissance sweeps in the nearby forest to prevent the anticipated enemy encroachment. On their second day at Rita, 2nd Platoon was on a morning sweep just beyond the wood line outside the perimeter. Spec4 Gary Hershley from Minnesota was walking point along a winding foot trail, armed with a 20-gauge pump shotgun. As Gary rounded a curve in the path, he abruptly came face to face with two North Vietnamese soldiers walking nonchalantly side by side in the opposite direction.

Gary was carrying his shotgun in the ready position, pointed down the trail. The two Communists had AK-47s. Unfortunately for them, they were carrying their weapons strapped over their shoulders as if they were on a long hike. The American soldier who was on high alert made short work of the two North Vietnamese who weren't.

In examining the two dead soldiers, it was clear from their uniforms and gear that they were not guerrillas, but main force soldiers. This finding was a clear indication that, as suspected, regular NVA forces were on the move in close proximity.

On 31 October 1968, Halloween night, events came to a head at the remote firebase. The senior officer present (and therefore the commanding officer) was Major Charles Rogers, 8-6th Field Artillery.

On three occasions that night, the base conducted a recon-by-fire technique known as a "mad minute". This called for all elements, at a prearranged time, to open fire on the surrounding tree line for three minutes. The idea was to discourage possible approaches by any nearby hostile forces. This tactic was rarely used since it gave away your position. In this case, the enemy clearly knew where we were.

The third "mad minute" took place at 0340 hours in the morning. The firing from the American positions ceased after three minutes as scheduled. However, to our dismay, the base was now receiving heavy incoming fire. The tactic had caught the enemy in the final stages of launching a ground attack on Rita—one of the most feared events of the war. A ground attack resembled the banzai attacks used by the Japanese in WWII—an all-out infantry assault intended to overrun and wipe out the defenders.

Heavy RPG fire was being taken in the first twenty minutes of the fight, and two American APCs were destroyed. An enemy element of approximately fifty soldiers took advantage of this weakened spot in the perimeter to breach the wire. Another RPG took out a self-propelled 155mm howitzer. Col. Rogers personally rallied personnel to launch a counter assault on the infiltrated NVA soldiers.

During the fight, Rogers was severely wounded in the face, yet continued to lead defenders to eliminate the hostile forces within the perimeter. Several times, he personally helped load the artillery tubes by setting the fuses.

The remainder of the perimeter brought heavily concentrated fire on the assaulting forces, including the use of "beehive" rounds—specially designed shells containing several thousand steel darts fired at point blank range into the oncoming NVA.

Meanwhile "Spooky" C-47 gunships circled the embattled base, pouring minigun fire into the attackers. Supporting artillery fire from nearby firebases Julie and Dot added to the frenzied defense. Air support from helicopter gunships completed the devastating firepower brought to bear in a desperate effort to save the base from being overrun.

The initial attack was beaten off, as was a second assault around

0500 hours. A final enemy effort was made at daybreak, but their energies had been spent. Sporadic fire continued throughout the morning, but the defense had held. It was later estimated that 1,300 rounds of indirect artillery fire, 50,000 rounds of 50 caliber machine gun, 500 rounds of 81mm mortar, and countless rounds of M-16 fire had been expended.

For his actions that night, COL Rogers was awarded the Medal of Honor. At that time, Rogers was the highest-ranking, black soldier to have received this ultimate medal for valor. Rogers spent many years in the Army, attaining the rank of Brigadier General. After retirement, he served as a pastor in his home state of Oklahoma.

Eighteen Americans were KIA during the battle. Twenty-seven NVA bodies were found inside the perimeter. Aerial recon of the area identified as many as 400 enemy bodies lying in the surrounding forest. Within a few days, the American forces shut down Rita, and Bravo Company was returned to our regular area of operations near Thunder Road (Highway 13) in Binh Duong province.

The abandonment of areas that had been fought at high cost was unfortunately typical in that war without front lines.

Years after the war, as a sort of hobby, I did a lot of research on the war, especially in regard to 1st Infantry Division operations. I collected copies of the tactical maps we used and meticulously researched the precise location of the different firebases we operated from, including, of course, Rita which surely must have been the biggest fight of Bravo Company's service during the war.

My hobby expanded to include research into other battles and points of military interest. One incident I found particularly interesting was the experience of film maker Oliver Stone who served in the 25th Infantry Division at approximately the same time I was with the 1st Infantry Division. Stone's experiences in

Vietnam were the basis of his Academy Award winning movie "Platoon." I try to give Hollywood's version of the war a wide berth, but I thought "Platoon" was one of the better efforts, due to Stone's first-hand familiarity with combat there.

The film's climax comes at the end with a dramatization of a ferocious ground attack on a firebase where Stone was one of the defending infantrymen in real life. I probed deeper into the history of that real fight in January of 1968. It was the assault of Firebase Burt, manned by Stone's 25th Infantry Division company. Since the 25th and the 1st Divisions' AOs (Area of Operations) were adjacent, I knew the FSB (Fire Support Base) Burt battle must have been near the area I served in.

I looked up the precise location of FSB Burt and was astonished to see it was the exact same location as Rita! I found a drawing of Burt and sure enough it looked exactly like a depiction of Rita, complete with the dirt track bisecting the base defenses. The Army had a habit of closing firebases when they were no longer needed for the current tactical plan, then reopening and remanning them when events made that location relevant again. I realized that Burt/Rita was just such a location—again reinforcing the counterintuitive US determination to fight ferociously for a spot of ground, only to abandon it, then fight again later for the same piece of geography.

In 2024, I returned to Tay Ninh province with the intent to find the site of Rita. The northern part of the province is still lightly populated and still pretty much the back of beyond. I hired a car and looked without success in the area where I knew the firebase once was located. Finally, I stopped at a local outdoor pool hall and asked the players in my limited Vietnamese if they could help. To my surprise, one of the young men told me he knew what I was looking for and offered to take me there.

After a short drive down a long, lonesome road, we pulled to the side, and he pointed out a primitive shrine back in the trees. The modest memorial was maintained by the local people, and as he told me, "Because of all the soldiers that died there." I took a couple of photos, said a short prayer, and headed back to Saigon. The journey across time and space had come full circle—thousands of miles and 56 years.

Here's a heartfelt parting salute to Gen. Rogers, to all my comrades in B Company, to the US soldiers of the other units who defended Rita, and to the enemy combatants who lost their lives on Halloween night 1968.

THOMAS A. "TOM" ROSS

ARE WE WHO WE WERE – BEFORE WE CAME?

"Good morning, Trung Uy (meaning Lieutenant)," Pop said when I walked into the team room.

Pop was our Vietnamese cook. He was a small man with large skills when it came to preparing meals for our Special Forces team. His culinary skills and repertoire had been developed and refined when he learned to cook for the French while they were in Vietnam. Pop was a warm, kind, and soft-spoken man whose cooking skills were only surpassed by his caring manner and the charm of his humble demeanor. Even though he was paid as a civilian employee, Pop had adopted the team and treated all of us as if we were his own sons. In turn, the team had adopted him and cared about him as if he were a close family member.

I was still rubbing my eyes and trying to remember everything I had to do that day.

"You want some breakfast?" Pop asked.

"No, not right now, Pop. I think I'll check on something first. Thanks, though. I'll eat a little later."

"Okay, Trung Uy. Tell me when you hungry."

Knowing that one of our other ambushes had contact with an enemy unit the night before, I knew there might be intelligence work that needed to be done early in the day. Certainly, before the sun rose too much higher and the day grew really hot. I walked over to the intelligence shack to see if Sergeant Koch and Bau, one of our interpreters, were there yet. When I reached the shack, both were already at work, waiting for me.

"Do we need to go to the river?" I asked.

"Yes, they got three last night. We should go now before the sun is on them too long," Sergeant Koch said.

"Yes, I know. Okay, let's go ahead and go now."

* * *

Whenever there was a contact in which VC (Viet Cong) or NVA (North Vietnamese Army) soldiers were killed, the common practice was for the camp's Vietnamese units to take the bodies down to the river near Dien Khanh village. There, the bodies would be lined up on the riverbank. Because many of the VC had grown up in local villages, their bodies were left along the river to be claimed by family members.

Members of the camp's Vietnamese intelligence network would be positioned at various points along the riverbank in an effort to determine the family names of those claiming bodies. This was done because it seemed reasonable to believe that those families might also be VC sympathizers and supporters. Any bodies that weren't claimed by sundown were buried.

I had been with our Vietnamese intelligence counterparts many times before to search the bodies for maps and any other information that might be of value. Normally, that unpleasant task was theirs, but this morning the job would be mine alone.

Approval had recently been given for a Korean mission into the Dong Bo Mountains, an enemy stronghold. The Koreans had three thousand men they could commit to the operation. Colonel Chang was the commander of a brigade-size unit of the Koreans' 9th White Horse Division. With his unit going into the Dong Bo, he asked that A-502 provide him with our most current information on any enemy activity in or near the mountains. Because the ambush the night before involved an enemy unit that had probably come from the mountains, I would personally conduct a search of those bodies.

When we reached the river, the bodies were laid out on display. They had already been lying in the sun for a while, so they were bloated—blown up like huge rubber balloons. The sight, even though I had seen it too many times before, was still a disgusting one. The bodies, riddled with bullet holes, bore testament to the battle that had cost them their lives.

The skull on one of the bodies had been shattered by the impact of the killing bullet. The skin that once covered the skull, including the face, now lay on the ground like a deflated volleyball. The skull of one of the other bodies retained its form but was open like a cracked egg and the brain was exposed to the morning sun, which was already becoming hot. Swarms of flies were gathering while others already present were crawling over the eyes and in and out of the nose, ears, and mouth of the three bodies. The sight was extremely repulsive.

I clearly understood the purpose and the intelligence value of placing the bodies by the river. But even though we were at war, it just didn't seem right that a human body, even a foe's, should be exposed to such abuse and disrespect.

As I gazed across the morbid sight, I thought, *Well, I may as well get this over with.*

When I approached the bodies to begin my search, a foul stench was already rising from the corpses. The nauseating odor was gagging—it was the smell of death. Knowing I wouldn't be able to hold my breath for the entire time it would take to complete the search, I took slow shallow breaths through my mouth, hoping to avoid the pungent smell. Unfortunately, that didn't help much. The putrid odor of death was much too strong.

Bending down over one of the bodies, I began to search the dead soldier's pockets for information. The body was so bloated that the worn khaki uniform had been drawn extremely tight around it. Even in death, the body resisted my intrusion. As I was going through the shirt pockets, I noticed a chain hanging around the corpse's neck. Pulling the chain out from under the man's shirt, it slid between my fingers until a medallion popped out and rested in the palm of my hand. It was a gold Saint Christopher's medal—very similar to one my grandmother had given me. At that moment, mine was hanging around my neck.

Probably Catholic, I thought.

I placed the medal back inside the soldier's shirt and told a nearby guard who had been posted to keep onlookers away to make sure the medal stayed exactly where it was until the family claimed the body or until it was buried. Continuing my search, I checked the rest of the soldier's pockets but found nothing.

While checking the pockets of the next body, I found the picture of a young woman. *Girlfriend? Wife? Sister?* I wondered. Whoever the young woman was, she didn't know what I knew at that moment. The young man who was carrying her picture . . . wouldn't be coming home alive. I imagined her grief when she was given the news. *God,* I thought, looking at the faded picture, *this body could easily be mine, and someone could be going through my pockets, finding pictures of my family.* Quickly dumping the thought, I continued to search the soldier's pockets for intelligence information.

In the next pocket, I found a small, folded piece of paper. It was a crudely drawn, but clearly detailed, map with Vietnamese writing on it. Potentially valuable, it appeared to be a map that may have guided the three soldiers to the village where they had been killed. The body searches were created to recover exactly this type of intelligence. I put the map in my pocket and would deliver it to the Vietnamese intelligence officers when we returned to camp.

A search of the third body revealed nothing.

* * *

With no conception of it at the time, I realized later, when writing about that day, the great irony in what had occurred that morning. These men may have lost a battle, but the army for which they had fought and given their lives would, in the end, be victorious. However, even with that eventual outcome of the war, by finding the map that day, their deaths would result in a similar fate for more than seven hundred of their comrades.

* * *

With my repugnant task complete, I stood and looked at the gathering of onlookers. In the crowd, I noticed an older Vietnamese woman (60s) approaching. She was slender, wearing a traditional peaked straw hat, and was carrying a round basket of fruit and vegetables under her arm.

The woman approached the bodies tentatively—then she shrieked, "Aahhhh!" She dropped her basket and wilted to her knees before one of the bodies, fruit and vegetables fell and rolled among the dead. Now, on her knees, the woman leaned over the body and cried loudly in Vietnamese, "My grandson! What have you done?!"

Obviously, the woman was his grandmother. She straightened up on her knees, back arched, arms outstretched toward the sky, and she screamed mournfully in Vietnamese, "You have become Viet Cong!"

Distraught, the woman fell across the swollen body of her dead grandson. I wanted to comfort the woman but doubted that she would welcome comfort from one of the soldiers responsible for her grandson's death, VC, or not.

Moved by the woman's agony, I did what I could. Turning to Bau, I pointed and said, "Tell that guard no one is to touch this soldier until his family comes for him."

Bau quickly translated to the guard who was standing only a few steps away and the guard immediately nodded acknowledgment.

The grandmother, lifted her face, wet with tears, and looked into my eyes. She nodded what I took as a painful gesture of appreciation, and said what I recognized to be "Thank you" in Vietnamese.

I turned and left, with Sergeant Koch and Bau in trail, to return to Trung Dung. As we walked away, locals hurried to the grandmother's side. I was glad she was being comforted.

During the short ride to camp, I thought about the task I had just completed and realized that as gruesome as the task was, it wasn't affecting me as intensely as it had when I first began the searches. In fact, being a part of such distasteful duty had become all too routine for me. After washing up and changing my clothes, despite what I had just seen and done — I was still able to sit down and eat a hearty breakfast.

As I sat at the table and ate the scrambled eggs Pop had prepared for me, I reconsidered some of a visiting British Colonel's questions regarding the war. While I had answered his questions and expressed opinions honestly, I had failed to mention to him one of my most important concerns: my fear that many of us may have been forever changed by service in Vietnam, and not in a good way. I conjectured—*Are we still who we were before we arrived in Vietnam? When we returned home, will we be the same persons we were before service to our country?*

Without question, we were all being changed in some manner by our experiences. I was saddened to think that I might be so hardened by those experiences that a girlfriend's touch, my mother's voice, or my sister's smile might not mean as much to me as they once had.

* * *

When I finally returned home to the States, I discovered that my wartime experiences had not affected me as I feared they might. To the contrary, I had a much greater appreciation for life, its beauty, and its gentle accompaniments. My mother's embrace and kiss on the cheek, my sister's tears and welcoming smile when I arrived at the airport in Pensacola, and my father's handshake and spontaneous bear hug when he saw me for the first time, meant more to me than they ever had before—not less.

However, many of those who served in Vietnam weren't as lucky as I was; they have never been the same. Some of them returned with their lives, but with a large portion of themselves, their essence, left behind in Southeast Asia. One of my teammates, over fifty years later, still takes meds in an effort to live a "near normal" life.

Fighting any war can be a horrific experience. Because of the divisiveness of the Vietnam War, some veterans continue to be saddled with immense guilt and anguish regarding their roles during the war. They have each had to bear the weight and psychological pain of their burden, essentially alone. Fortunately, I am not one of those veterans, so it is easier for me to make relatively unbiased observations. One such observation is that it was all too easy for some people to blame the soldiers for horrible things that happened as a result of poor decisions made by those at the highest levels of our government. Our troops do not create wars—they fight them.

As true as any statement that can be made about my experiences during the Vietnam War is the fact that virtually every American man or woman that I met was trying to do a good job — patriots who simply wanted to serve his or her country. The Vietnam War is considered by many to be a failure in every way. However, any failure is not the fault of those who were sent to fight it.

It has been said that American forces never lost a major battle in Vietnam. Unfortunately, that isn't true. We experienced many losses that have been well documented. But what is true is that we won a great many more battles than the few that were lost.

Regardless of wins or losses, as time has passed and their numbers thinned, some of the men and women who served with great honor and dedication may still be trying to be who they were before they went to Southeast Asia—may God bless them all.

PETER G. MANFRE

BRANCH OF MILITARY AND JOB HELD: US Marine Corps, Aerial door gunner on a CH-46 helicopter
DATES OF MILITARY SERVICE: April, 1968 - June, 1972
UNIT(S) SERVED WITH IN VIETNAM: 9th Marine Division, 1st Marine Air Wing, Marine Medium Helicopter Squadron 265
DATES SERVED IN VIETNAM: March, 1969 - September, 1969
HIGHEST RANK HELD: E4
PLACE AND YEAR OF BIRTH: Newark, NJ 1946

RECEIVED TWO PURPLE HEARTS

Peter G. Manfre of Cumming, GA, a double-Purple Heart recipient, volunteered to serve his country on April 10, 1968, at the height of the Vietnam War by enlisting in the United States Marine Corps. He was ordered to Vietnam in March of 1969 and assigned to the 9th Marine Division, 1st Marine Air Wing, Marine Medium Helicopter Squadron 265. He volunteered to be an Air Crew member (aerial door gunner) on a CH-46 helicopter.

First Purple Heart: On July 2, 1969, after receiving enemy gunfire in Quang Tri Province, his helicopter was shot down. Peter was knocked unconscious and suffered severe head trauma after the helicopter fell approximately 1,700 feet, crashing to the ground and bursting into flames. When Peter woke up, all he saw was the helicopter was on fire, and he was strapped in a harness and could not get out. The impact was so great, with a blow to his head, that his flight helmet was cracked right down the middle and both of his knees were injured. Peter was rescued by his crew chief who had to cut him out of his harness. At that time, he was unaware of several back (disc problems) that were uncovered by the Veterans Administration (VA).

At his debriefing, they were amazed he was still alive. Bullets had come through alongside his gun mount that had blown out the hydraulic fuel main line to both engines. He had moved up to look over the ramp for enemy muzzle flashes. Had he been sitting down at his gun mount, he would have been another statistic. Peter considers that to be his luckiest day ever.

Second Purple Heart: Not a Marine to be dissuaded after being confined to his rack for 24 hours, the next day he went right back up and requested another mission. Two months later on September 9, 1969, while on another troop movement mission, his chopper was approaching an LZ (Landing Zone) that had enemy activity and was ambushed. He was hit by enemy small arms fire and received severe gunshot wounds in his left ankle and right arm. The helicopter pilot was also wounded.

Peter said, "This was like a three-thousand-pound steel door slamming into my face. Laying on the floor of the helicopter, I saw my right forearm muscle hanging outside of my flight suit. As I was being bandaged up, they took off my left boot. Normally, when standing, your leg and foot are "L" shaped. Laying on my back, when they took off my boot, my foot fell backwards the other way.

"The flight out to our ship was nothing short of more unknowns. The helicopter was shot up and we had a bumpy, bumpy ride out to the ship. I was lying on the floor and could see that we were flying just above the waves in the ocean. Not good, we could go down in the drink (ocean) at any time.

"The next thing that I saw was the co-pilot making a left turn where I could see the sailors on the ship's deck wearing their foam white suits and their hoses at the ready. I said to myself, 'I really don't need this, now.' Fortunately, the helicopter landed without an incident."

Peter was hospitalized aboard the USS Iwo Jima and medevacked to hospitals in Japan and Guam. Due to the severity of his injuries, he was further medevacked back to the States.

He spent almost one year of his life in rehabilitation recovery in the US Philadelphia Naval hospital, learning how to walk and use his right hand again, then spent at least another year in ambulatory recovery at home.

From there, Peter was placed on Permanent Disability and given an honorable retirement from the Marine Corps on June 1, 1972 due to combat wounds.

As a reminder of his wounds, later in life Peter developed ongoing vision, hand, arm, and back troubles requiring four brain surgeries—along with Agent Orange (presumptive) diabetes and prostate cancer.

Peter Manfre had volunteered to serve. Even though he could

no longer wear the cloth of his country, he never stopped serving. Peter currently not only holds memberships in the following veteran organizations, but is active in a number of them, holding or having held leadership and board positions and lifetime memberships in many:

American Legion Post 201 (LF)
Atlanta Vietnam Veterans Business Association (LF)
Blinded Veterans Association (LF)
Disabled American Veterans (LF)
Johns Creek Veterans Association
Johns Creek Veterans Association Color Guard
Military Order of the Purple Heart Chapter 576 (LF)
VetBuds
Veterans of Foreign Wars (LF)
11th Hour Squadron
(LF = Life Member)

Additionally, he is a Fourth Degree, Sir Knight with over 45 years of service and Honorary Life Member of St. Benedict Catholic Church Knights of Columbus (KofC) and is the interface for several years that helps operate an annual golf tournament where proceeds go to veteran organizations.

Peter is consumed with compassion for veterans with health issues and spends untold hours a week on the phone checking in on KofC and other buddy checks on veterans who have shared some of the same medical issues as him.

Here is a message from one of Peter's fellow Vietnam Veterans: "Peter Manfre volunteered to serve. North Vietnamese bullets could not keep him from that calling. When we look up the definition of 'service,' in the dictionary, Peter Manfre's picture should be right there."

You may ask, "But what happened to Peter Manfre after serving our country? Did he have a career? After hearing all of this, we have good news: here is a short career update.

Peter was honorably retired from the U.S. Marine Corps on 1 June 1972, after four years and two months of service — that is 17 months of active duty in the Marine Corps.

The majority of his professional life was spent with Bell Laboratories and AT&T in New Jersey and IBM in Georgia. He retired from AT&T after 26 years and two months. He also retired from IBM after 16 years of service.

RICHARD H. MOUSHEGIAN

BRANCH OF SERVICE AND JOB: US Army — Quartermaster Officer

DATES OF MILITARY SERVICE: Active Duty: 1968 to 1980; Reserve Duty: 1981-1993 (24 total years)

UNIT SERVED WITH IN VIETNAM: HHC, Long Binh Post, USARV

DATES IN VIETNAM: Oct 1968 to Oct 1969

HIGHEST RANK HELD: Lieutenant Colonel (O-5)

PLACE AND YEAR OF BIRTH: Camp Wheeler (deactivated after WWII; near Macon), GA — 1943

TWO BROTHERS IN VIETNAM

Brother Stephen (Steve) Moushegian (1st Lieutenant) and I were in Vietnam at the same time. He was an Armor officer that had just learned to fly the army's helicopters, and I was a Quartermaster (logistics: supply and maintenance) Captain. I was in Vietnam first and had no intention of leaving until I completed my Vietnam tour. He came later after graduating from the Army's helicopter schools (Fort Walters at Mineral Wells, TX, and Hunter Army Airfield at Savannah, GA) knowing full well that he would be assigned to Vietnam.

Since we were brothers, I requested that he be assigned near Long Binh Post where I was. Permission was granted, and he was assigned to the II Field Force (Aviation), also at Long Binh Post, with a mission of providing air transportation for high-ranking military and civilian personnel using Huey (UH-1) and Loach (OH-6A) helicopters.

This story is what Steve relayed to me about some of his aviation experiences in Vietnam. Since I was also an Army officer, I could place his experiences in military context. Since I became a pilot after retiring from the army, I could understand a great deal of the aviation terminology, practices, and situations, but not all the military-aviation practices.

First, I must provide an explanation of Steve's personality: Besides being smart, interested in military history, and excellent at assessing people and circumstances, he loved conflict with people, was "rough around the edges" (probably *not* a description of a touchy-feely person with warm, personal relationships), and always wanted things done his way—perfect for leadership—but tough when he is not in charge.

Additionally, the military loved him in a combat zone (Vietnam) because he was *highly focused and fully committed* on important tasks (which explains his medals of valor: Silver Star and four Distinguished Flying Crosses) when he later volunteered for an in-country transfer so that he could fly "real missions" with the 11th Armored Cavalry Regiment (ACR) as an aerial scout.

When Steve arrived at his first aviation assignment with the II

Field Force, they assigned him a "hootch" (living quarters with two beds, two wall lockers, and one desk with a lamp), and he realized that he had a roommate whose personal belongings were already in the room. What he did not realize was that his roommate was another officer-pilot that loved conflict with people, was just as "rough around the edges," and wanted things done his way—just like Steve.

After in-processing and eating supper, Steve returned to the hootch to be greeted by his fuming roommate who was highly irritated with the intrusion because he had made it perfectly clear to the operations and billeting people that he would only settle for a private room—no roommate! (Keep in mind that the other *Lieutenant* had *demanded* a private room in the military!)

The initial greeting/encounter literally began as a shouting match between the two of them. (The roommate I will call Burns because I'm not sure if that was his name since Steve passed away in 3/22/2019 and was interred in Arlington National Cemetery on 4/17/2020.)

Burns yelled at Steve to get out of his room, and Steve shouted back, "I can't. This is where I was assigned. Besides, YOU go tell them!" Gradually, they realized that they were carbon copies of each other—both being mavericks. So, they continued to communicate with each other by shouting, but they would not let *anyone else shout at the other*; if so, the outburst was immediately challenged with a "Back off!" with no further explanation given. As you can probably guess, they became strong friends, even though no one else understood them.

Before Burns rotated (i.e., was reassigned) back to the States two to three months later, he told Steve, "I'm going to tell you some things that no one else will tell you." (Notice that Burns used the word "tell" not "share," since neither one of them was touchy-feely.)

Item 1: Burns pulled out a very well-worn scrap of paper from his wallet and said: "Copy this Vietnamese address *exactly* as I have written it and keep it in your wallet. When you fly to Saigon and wander around the city for whatever the reason, this address is the *only* address that the taxi driver will understand. He will not understand your explanation that you want to go to the American helipad near the 'big PX (Post Exchange).'" Consequently, Steve used his piece of paper on several occasions, and the taxi driver took him exactly to the right location, every time.

Item 2: "When you take a taxi (in Saigon), only pay the *going rate that Vietnamese pay*." (Burns briefed him on the appropriate pricing structure.) He went on to explain that after you exit the taxi, offer the money. If the driver protests, throw it through the window onto the passenger seat, and kick the side of the vehicle (just loud enough for him to hear) as you register your protest at the attempt of price gouging. The driver will drive away in a huff.

On one occasion, Steve and I were in Saigon, and, at the conclusion of the trip, Steve asked me to get out and stand on the curb. When he offered the money, the driver protested, and I immediately offered more money to settle the dispute, but Steve said, "Stay on the curb. I'm handling this." Then, Steve proceeded with the Burns' protocol.

Item 3: Everyone knows that the US Navy serves excellent food. Nabay (pronounced "Nah-bay"; actually, Nhà Bè) was a US Naval Support Activity base with helipad that was located about 10 miles SSW of Saigon at the junction of two rivers very near the South China Sea. Burns explained that if you are "flying in the area" of Nabay and want to get an outstanding lunch with the Navy, choose a fake call-sign when coming in for a landing. That way, any record of your call-sign by the tower will NOT be linked back to your unit and will not precipitate in a demand for an explanation of why you were there. (Tower's response to a possible query from your unit: "No. No one with your unit call-sign landed here."—Mission accomplished: excellent lunch and no tracing back to home unit!)

Item 4: The last piece of advice from his roommate involved the approach to Tan Son Nhut Air Base (near Saigon) by helicopter, an air base that served the US Air Force, Vietnamese Air Force, US civilian charters, and Vietnamese commercial aircraft. The base was extremely busy with fixed-wing aircraft landing and departing every few minutes on two, parallel runways. The normal procedure, according to Steve, after entering the airport airspace is to come to a hover near the runways and wait for a break in a continuous stream of radio transmissions between the airport tower and all of the arriving and departing airplanes in order to advise the tower that you are *crossing the active runways* to land on the helipad on the other side. If the helicopter pilot hesitates or stammers in his announcement, asks a question (weather update/advisory?), or wants to chit-chat, the tower will immediately demand that he leave the highly congested, airport airspace—*regardless* of the helicopter's mission!

Burns' procedure was very simple, but it took *nerves of steel:* (1) come to a hover and *wait until someone takes a breath* over the radio, (2) *immediately key the mic* (radio microphone) and quickly proceed with your announcement that you are crossing the runways to land at the helipad, (3) and, **without waiting for a response from the tower**, *pitch the aircraft over in full acceleration to fly in between the departing airplanes* and proceed smartly to the helipad. Steve found that the Burns' procedure was extremely helpful on rare occasions of flying official missions onto the air base.

The true test of Burn's fourth piece of advice came after Steve volunteered for an in-country transfer to the combat unit, the 11th ACR out of Quan Loi. He had been flying for the organization for several months and was highly respected for his piloting and combat skills when his CO (Commanding Officer, a Colonel in this case) mentioned that he needed to go to Tan Son Nhut Air Base.

Steve asked, "Sir, how are you going?"

The answer, "By helicopter. Why?"

Steve: "You will never make it."

Colonel: "Why not? My pilot (a very experienced Warrant Officer, probably a WO-3) knows where it is."

Steve: "The tower will never let you in."

Colonel: "What do you suggest?

Steve: "Sir, for that mission, I will fly with you as your co-pilot and your WO as pilot, and I will *make sure* that you get into Tan San Nhut—I've been there several times already. (Keep in mind that this offered arrangement was *highly unusual* for Steve in that he *hated being co-pilot!* However, what he loved even more was a *good challenge!*)

"However," Steve continued, "when we *approach* the air base and we are *in the airport airspace*, your pilot must follow my directions—immediately and to the letter—and I will be making all the radio calls. If you don't approve, they will never let us in."

Colonel: "Approved."

Steve: "Yes, Sir. I will brief your pilot."

When it came time to fly the mission to Tan Son Nhut from Quan Loi (about 60 miles away), Steve made the radio calls, waited for someone to take a breath over the radio, and quickly made the appropriate announcement. Then, Steve pointed to a break in the aircrafts taking-off and told the pilot to "Hit it!" The WO pilot pitched the helicopter over in full acceleration, navigated between two C-130's taking off (under full power), and quickly landed at the helipad on the other side of the runway.

Colonel: "Damn fine job."

The egress at the conclusion of the visit was equally as harrowing, but was without incident.

(Footnote for clarification: By the helicopter making the announcement to the airport air traffic of his intentions, the aircraft on take-off roll did not need to abort take-off, and the aircraft in the air on climb-out did not need to take evasive action because the helicopter off to the side would go behind him to cross the runways.)

JAMES J. "JIM" HOOGERWERF

BRANCH OF SERVICE: USAF — C-130 Pilot
DATES OF MILITARY SERVICE: 1966-1973
DATES SERVED IN VIETNAM: May 1968 - May 1970 (not inclusive)
UNITS SERVED WITH IN VIETNAM: I was permanently based offshore in the Philippines at Clark Air Base 773rd TAS, 463rd TAW. In country I was TDY to 834th Air Division, Det. 1 TSN (1968-69), Det.2 CRB (1969-1970)
HIGHEST RANK HELD: Captain (O-3)
PLACE AND DATE OF BIRTH: Detroit, MI, 1943

EVERY PILOT A FIGHTER PILOT

It is a privilege to have experienced the intensive training of a USAF student pilot. On December 6, 1966, I reported to Webb Air Force Base, Big Spring, Texas; I was 22 years old, and a newly minted Second Lieutenant. Students progressed from the Cessna T-41 (civilian 172) and twin-jet Cessna T-37 to the supersonic T-38 Talon advanced pilot trainer. If all went well, in fifty-three weeks I would pin on USAF silver wings.

Instructors challenged students with the mantra "every pilot a fighter pilot." However, motivation was not an issue; strapping on the T-38, you had to be at the top of your game. That attitude carried over to our assigned aircraft after flight school: F-100, F-4, F-111, B-52, KC-135, EC-47, C-130, etc. My assignment was the Lockheed C-130B at Clark Air Base in the Philippines. It wasn't a fighter, but its capabilities allowed it to be flown like one.

After fifty-three weeks, my wings were assured. All that remained was the pinning. Unexpectedly, I had a last opportunity to fly the T-38 "like a fighter pilot." I will never forget it — it was a solo flight — no instructor (normally in the rear cockpit). The situation was unusual: I may have had some extra time to log. Whatever, a supersonic high-performance jet was at my disposal. What a gift!

Operations assigned a high-altitude training area (where jet engines are more efficient). This was important to conserve fuel because my flight was planned for 1.4 hours, the maximum endurance with full tanks. I decided to do an afterburner climb. This was not part of training and I had not flown one myself, but I sat through a couple with instructors, so I knew the procedure: accelerate to 300 knots, level at 10,000', get clearance from ATC (Air Traffic Control) for an unrestricted climb, light the burners (two of them), and pull the stick back to initiate a steep, nose-high attitude. The burners suck fuel at a high rate, which would seem counterintuitive for saving gas, but it would only take about a minute to get up to my area with a climb rate of 30,000 fpm (feet/per/minute).

Everything went as planned… initially… but, as I rocketed skyward, I took a peek outside… off the gauges… mistake! By the time I focused my attention back in the cockpit, I hurled out the top of my area — climbing — with no thought of leveling off. I had to get down fast! Simultaneously, I closed the throttles, rolled inverted, and sucked the nose down below the horizon. While doing this, I topped 50,000' (above 50,000' a pressure suit is supposed to be worn).

Not quite having a grip on things and rapidly descending, I rolled right-side-up and raised the nose to stop my descent. By then, I had busted the bottom (altitude) of my area! Note: Leaving your area was a serious infringement and would result in an automatic "pink" (failure) by an instructor... I had busted both the top and bottom! Fortunately, I was solo, and I wasn't going to tell on myself. ("Fighter" pilots aren't stupid!)

Gingerly, I climbed back into my assigned airspace, cruised around for a while, then headed back to Webb. Since this was an unsupervised flight, there was no debriefing (i.e., evaluation) required, but in assessing my performance and since few fighter slots were available, I had made the right choice in selecting the C-130s.

A few days later, Mom pinned on my wings, and in five months I was flying C-130s in Vietnam (May 1968 — May 1970).

THERE IS MORE TO MY T-38 STORY, "EVERY PILOT A FIGHTER PILOT"

My frame of mind for becoming a pilot never leaned toward fighters. Air travel growing up influenced me — I liked big airplanes. They took me to interesting places. Choosing the C-130 was never in doubt for me, but I played along with the fighter mentality — the T-38 was really cool. Few fighter slots were available anyway.

My only option would have been the back seat of the F-4 Phantom. The C-130 was built for combat; it was designed with fail-safe systems built-in. It was assigned missions under the Tactical Air Command (TAC), but staged from Clark Air Base in the Philippines. My tour was twice as long as being permanently assigned in-country. However, time spent at Clark would be a nice break from flying combat.

There were other reasons not to choose fighters as well: In July 1966, North Vietnam paraded captured pilots through Hanoi streets as throngs of angry people beat them, and the communists threatened to try them as war criminals. I had a premonition that would be my fate if I flew the F-4. You flew 100 missions over the North; then, you could upgrade to the front seat (as pilot) to fly another 100 missions… no, thanks!

More about my last T-38 flight — getting "pinked" was possible. Although I was solo, I suspected a two-ship formation was operating below me. I might be seen in their airspace.

Another concern came from the ATC controller asking to report my altitude. Dispatchers (operations) randomly monitored frequencies so I could not admit I was outside my authorized airspace. He asked several times, and I finally found out why… he pointed out a B-52 passing overhead (well clear).

Well, with these concerns, I walked by the classroom door to glance in and see if anyone was waiting for me. No one was, so I quickly went in and logged my flight.

Thus ended my brief saga as a "fighter" pilot, although not my deep respect and admiration of fighter pilots and that culture.

WILLIAM H. (BILL) BROWN

BRANCH OF SERVICE: United States Army — Signal Corps Officer
DATES OF MILITARY SERVICE: August 1966 — August 1974
VIETNAM SERVICE: Platoon Leader, 2nd Platoon, B Company, 121st
Signal Battalion, 1st Infantry Division (5 months), 1st Infantry Division
Combat Photographer Officer, 1st ID Combat Photographer Unit, C
Company, 121st Signal Battalion, 1st Infantry Division (7 months).
DATES IN VIETNAM: August 1967 to August 1968
HIGHEST RANK HELD: Captain (O-3)
PLACE AND YEAR OF BIRTH: Savannah, GA — 1944

SOME GAVE ALL — FRIENDS WHO GAVE ALL

I graduated from North Georgia College (NGC) in June of 1966. North Georgia, now the University of North Georgia, is one of six Senior Military Colleges authorized and recognized by Congress years ago. Many of my friends, alumni, classmates, and I found ourselves serving in Vietnam a year or so after graduation.

1968 was the year of the Tet Offensive, the most intense fighting of the entire war. God was with me in a special way during that time. With His help and protection, I didn't lose a single man from my platoon. However, I lost fifteen friends who gave their all serving in Vietnam, fighting Communism, and fighting for our freedom. Their names are engraved on the Wall at the Vietnam Veterans Memorial in Washington D.C. along with more than 58,000 others who made the ultimate sacrifice in that war. Here are my thoughts and memories about two of these honorable men, my dear friends.

1LT JOSEPH HILLMAN, III

This is the story of 1LT Joseph Hillman, Killed in Action at age

23, leading his men in combat in the Cu Chi area of the Republic of Vietnam in July of 1968.

Joe and I met and grew to be friends during our four years at North Georgia College (NGC). Joe was strongly committed to becoming the best Army Infantry Officer possible. At NGC, Joe took his military training and duties seriously, and he looked forward to serving his country in the United States Army as an Infantry Platoon Leader, a position of great responsibility and considerable risk.

During our senior year at NGC, Joe and I double dated on a couple of occasions in my '56 Chevy. We also worked out in preparation for representing our respective cadet companies in the push-up competition on Field Day.

A year and a few months after graduation from NGC, we found ourselves serving in Vietnam. I arrived in RVN on August 5, 1967 where I served as Platoon Leader, 2nd Platoon, B Company, 121st Signal Battalion, 1st Infantry Division, and later as the 1st Infantry Division Combat Photographer Officer. Joe began his tour of duty in Vietnam on December 3, 1967, where he served as a Rifle Platoon Leader in the 3rd Brigade of the 101st Airborne Division.

While serving in Vietnam, Joe earned *two* Silver Stars for Valor, a Bronze Star with "V" device (for Valor), two Purple Hearts for wounds received in combat, Vietnam Service Medal, Vietnam Campaign Medal, Army PUC (Presidential Unit Citation), Combat Infantryman Badge, Parachutist Badge, and Marksmanship Badge.

An excerpt from the book *All Gave Some, Some Gave All* states, "Joe led his men by example, and his menacing attitude kept him on the frontlines of combat duty. He spent a lot of time fighting in Trang Bang in the Cu Chi area, a city made infamous for its vast and complex tunnel systems that aided the Vietnamese tremendously in causing significant U.S. casualties." On July 21, 1968, Joe's company participated in a six-company air assault to place a cordon around a cluster of villages. The next day, on July 22, Lieutenant Joseph Hillman lost his life by a single AK47 round to the forehead.

Joe's NGC classmate and friend, Ralph Colley said, "The Joe we knew was a man of honor, a patriot by choice and conviction, fearless in conduct, and passionate about his life's interest."

In March of 2023, I nominated my friend, Joseph Hillman III, for induction into the Georgia Military Veterans' Hall of Fame. A copy of my cover letter, which includes excerpts from General Orders 808 awarding Lieutenant Joseph Hillman his first Silver Star, and the Silver Star Citation for Lieutenant Hillman's second Silver Star reads as follows:

From: CPT William (Bill) H. Brown, Nominator

To: Georgia Military Veterans' Hall of Fame (GMVHOF), 5805 State Bridge Road, Ste G 379, Johns Creek, GA 30097

Subj: Nomination of First Lieutenant Joseph Hillman III, U.S. Army (Infantry) in consideration for induction (Posthumously) into the Georgia Military Veterans' Hall of Fame (GMVHOF)

This is the formal nomination of First Lieutenant Joseph Hillman III, U.S. Army (Infantry) for induction (Posthumously) into the Georgia Military Veterans' Hall of Fame (GMVHOF). First Lieutenant Hillman is being nominated for the Valor Category for having served with valor and being killed in action during the Vietnam War.

Lieutenant Hillman was born in Marietta, Georgia on November 11, 1944. He graduated from North Georgia College in 1966 and was commissioned a 2nd Lieutenant. He arrived in South Vietnam in December 1967 and served as a Platoon Leader in Company B, 2d Battalion (Airborne), 506th Infantry, 101st Airborne Division. He was awarded two Silver Stars, Bronze Star with "V" device, two Purple Heart Medals, Vietnam Service Medal, Vietnam Campaign Medal, Army PUC, Combat Infantryman Badge, Parachutists Badge, and Marksmanship Badge.

General Orders Number 808 awarding his first Silver Star provide the following reason:

For gallantry in action in the Republic of Vietnam on 16 January 1968. Lieutenant Hillman distinguished himself while on a combat operation in Tan Yuen Province, Republic of Vietnam. As the patrol he was leading moved through thick jungle undergrowth, it encountered intense automatic weapons and machine gun fire. Although wounded, with loss of vision in one eye due to excessive bleeding, Lieutenant Hillman maneuvered his platoon and directed their fire while constantly exposed to the hostile fire. Locating an M-79, he moved into a firing lane in an attempt to recover a wounded man and silenced an automatic weapons position. He then moved across the firing lane under fire to a machine gun which he directed on the enemy positions, which allowed some of the wounded to be recovered from open areas. The outstanding display of bravery by Lieutenant Hillman was a source of encouragement for all the men and contributed immeasurably to the success of the mission. Second Lieutenant Hillman's extraordinary heroism and devotion to duty in close combat were in keeping with the highest traditions of the military service and reflect great credit upon himself, his unit, and the United States Army.

The Silver Star Citation for Lieutenant Hillman's second Silver Star reads as follows:

SYNOPSIS: First Lieutenant (Infantry) Joseph Hillman, III (ASN: 0-5333143), United States Army, was awarded the Silver Star (Posthumously) for conspicuous gallantry and intrepidity in connection with military operations against the enemy while serving with Company B, 2d Battalion, 506th Infantry Regiment, 101st Airborne Division, in the Republic of Vietnam. KIA 7/22/1968. **A Brave Platoon Leader.**

James Stevens Roach, colonel (retired), US Army, provided the following comment:

Joe Hillman was the best Platoon Leader in B Company, 2nd Battalion, 506th Infantry. He was smart, hardworking, and took care of his platoon. When there was a fire fight, he was always in the thick of it, looking for tactical advantage and taking care of his men. Joe was killed leading an assault against a Viet Cong stronghold. The company had been moving toward a night defensive position when they were taken under fire from this Viet Cong entrenched position. In that type of tactical situation, you either move quickly, or die, and Joe was the B Company leader that was first into the Viet Cong position. B Company won the firefight, but lost several good soldiers, and the best Platoon Leader in the Company—Joe Hillman. He is a real-life hero to all the survivors of B Company who knew him. Rest Well, Friend.

Ranger Roach
4th Platoon Leader, B Company, 506th Infantry
Oct 6, 2007

Respectively submitted:
CPT William (Bill) H. Brown
Platoon Leader, 1st Infantry Division
Vietnam — Aug '67-Aug '68
North Georgia College — Class of '66
Nominator for Joseph Hillman III
Member, AVVBA and GVVA

Joe Hillman was an NGC classmate, a fellow Vietnam Veteran, and a good friend. I will always remember and be thankful for our shared times at NGC. I appreciate his service to our country and his sacrifice for our freedom. He paid the ultimate price. He gave his all. He is worthy of my respect and admiration, and he has it.

1LT JERRY HALEY

This is the story of 1LT Jerry Haley, KIA February 3, 1968, defending the communications outpost on Nui Ba Den, the Black Virgin Mountain, during the early days of the Tet Offensive of 1968.

First Lieutenant Jerry Rankin Haley served with Company A, 121st Signal Battalion, 1st Infantry Division, United States Army, Vietnam.

My good friend, 1st Lieutenant Jerry Haley was killed during the early days of the Tet Offensive, February 3, 1968. He was a fellow Platoon Leader in my battalion, the 121st Signal Battalion, 1st Infantry Division. Jerry was three years older than me. He was married and had children. He was from Bastrop, Louisiana. As best I remember, he came up through the ranks, went to OCS, and earned his commission. He was an excellent officer, committed to doing his job well, and taking care of his men. He took our 1st

Infantry Division motto, "No mission too difficult, no sacrifice too great, duty first" very seriously.

Tet is the biggest celebration of the year for the Vietnamese people. It's a family reunion, a spring festival, a national holiday, and everyone's birthday. The Vietnamese people had been at war for many years, first with the French, and now with the North Vietnamese Communists. Over the years, Tet had become an unofficial cease fire. Many soldiers went home to spend the holidays with their families. They would return to continue fighting the war after the holidays.

Tet 1968 was different. The Viet Cong had planned a surprise attack at strategic locations all over South Vietnam. Late on the first night of Tet, the first thunderous barrage of rockets and mortars slammed into 1st Division Headquarters in Lai Khe base camp where my platoon and I were located at the time. Thus began the Viet Cong's brutal Tet offensive. Jerry and his platoon were located on Nui Ba Den, the highest mountain in III Corps. They were operating and protecting a critical radio communications site. Nui Ba Den was hit with a rocket barrage on February 3, 1968. The radio communication site was destroyed. There were casualties, and my friend, 1st Lieutenant Jerry Haley was one of them.

Not only did I lose a good friend, but I was the officer assigned the task of preparing his personal belongings to be sent home to his family. It was a heart-breaking experience. I could just imagine his wife, children, and parents going through his personal belongings, and how they would feel as they saw and held each item, knowing they would never see Jerry alive on this earth again, knowing they would never hear his voice, never hug or hold him. Jerry is buried at Charleston Cemetery between Stanton and Covington, TN.

Jerry Haley was a good friend. I'm thankful for the time we

spent together in Lai Khe. I was blessed to know him, and I'm thankful for the good memories. He made the ultimate sacrifice fighting for the freedom of the South Vietnamese people, fighting communism, fighting for our freedom. He gave his all. He deserves my respect and admiration, and he has it.

Respectively submitted:
CPT William (Bill) H. Brown

HUGH D. PENN

BRANCH OF SERVICE: 633 Combat Support Group, Pleiku Air Base (AB), Republic of South Vietnam
DATES OF SERVICE: Feb 1966 – Jan 1970
DATES SERVED IN VIETNAM: Feb 1969 – Jan 1970
HIGHEST RANK HELD: Staff Sergeant (E-6)
AIR FORCE SPECIALTY CODE: 73250A - Personnel Specialist

VIETNAM IN A "MACHINE ROOM"

I was a personnel specialist assigned to what was called the "machine room." Our task was to transfer an airman's personal information from written paper documents to IBM cards for entry into computers. We coded, we decoded, we sorted, we filed these cards using IBM data card machines—the IBM 024, IBM 026, and IBM 086. The job required a humidity-controlled environment to keep the cards from warping and jamming the machines. That meant our office had to be air conditioned. This was the daily routine of my job both stateside (the World) and in Vietnam.

I received orders for Vietnam in the fall of 1968. I was to report to Pleiku AB in II Corps in the central highlands, about 25 or 30 miles from the Cambodia/Laos border. I was going to the 633 Combat Support Group. Its combat function was to support air operations for the ground forces such as reconnaissance, interdiction, search and rescue, and any general operations to further the war effort. While the base served that vital function, my role was to maintain airmen records and to prepare packages of these cards to be transmitted to major air command for input into computers.

I arrived at Cam Ranh Bay in February 1969 from McChord AFB, Washington with a brief stop in Japan. After an orientation on Vietnam, I was issued jungle fatigues and boots, and we took a flight on a C-123 to Pleiku AB. I left the U.S.A. in the middle of winter, and less than a day later, I was in a tropical sauna. The heat and humidity of S.E. Asia is legendary. It was my good fortune that my role required an air-conditioned office. Even better, I found out I would be working with a good friend from Amarillo AFB, Texas.

The first week or so was getting used to life and wondering if all the booming noises were incoming rounds. That became clear one morning when an incoming 122 mm rocket educated me as to the difference between incoming and outgoing rounds. I was getting ready for a shower when it hit, and I raced to a bunker trailing a towel behind me. Life was routine—work until 9 pm, pull KP, fill sandbags, and watch the gunships (Snoopy and Cobras) work on the VC in the distance.

Going to a war zone is like no other experience a person will ever go through. Having a child, getting married, discovering Christ are some things in life that may leave everlasting memories, but for me, all other things pale in comparison to life in a war zone.

I had to adjust to the requirements of the job that were different from what they were Stateside. I also had to re-qualify on the M-16, marksmanship, and the assembly/disassembly and cleaning of the weapon. In basic training, I just barely qualified, but once I got orders for Vietnam, I shot Expert. It's amazing what a little motivation will do for you. (The military is great for motivation.)

I was very fortunate to be in the Air Force as compared to the Army or the Marines, and I knew and appreciated that branch of service. Funny, Stateside I had risen high enough in rank that I didn't have to pull KP (kitchen police), but at Pleiku I had to do that and my regular job. I was also assigned to fill sandbags used in building the defensive bunkers in case of an attack. We were trained on walking a patrol in the event we ever had to go outside the wire and look for the bad guys. With those things out of the way, I settled down to the day-to-day life in my war.

Some things stand out clearly, others have faded. In my mind, some of the sequence of events is muddled. Some of the men I worked with have faded from memory, some I will remember until I die. One guy, Sgt Whitehead from Ohio, was our scrounger. He wasn't in the office much, and how that happened, I don't know. Sgt Whitehead was our *go to guy* for things on the black market or unauthorized stuff.

He once got me a grease gun: a Thompson M3, submachine gun, 45 caliber. I didn't keep it long because there was nowhere to readily get 45-caliber ammo if you needed it. He once came up with an Army jeep, and how the heck he expected anyone could keep it is anybody's guess. He could always come up with steaks for Saturday cookouts on grills made from 55-gallon drums.

In retrospect, I now know that some of those drums had once contained Agent Orange. The same Dow Chemical product that has brought so much death to both Americans and Vietnamese. I have lost many friends to Agent Orange (AO). For years, I was sure I had no contact with it, but later, I learned that it was used extensively in the defoliation around the perimeter to allow a field of fire for base security. The runoff of AO into the lake where we got drinking water, the 55-gallon ½ drum grills—they were all contaminated with AO. The danger of exposure to Agent Orange was known by Dow at the time, but not by the military. It was a terrible deception to pull over the world.

Rocket attacks were the primary danger that came our way, and they could come at the most unexpected times: when you were sleeping, going to shower, going to chow, working, or watching a movie at the outdoor movie during a monsoon downpour. There was only one night when I had to get my M-16 and report to an assigned defensive bunker. There were many times when Air Police and augmentees were alerted, but the whole base generally wasn't. Art Zachary from Syracuse, NY was one of the guys in my office who was an augmentee for base security.

Another colorful airman I worked with was Roger Hurley from Kansas City, MO. SSgt Hurley was a swarthy, plump kinda guy. Most people would say he was not handsome, and that was being generous. He spent most of his career in the Far East, and it took us forever to understand why. Wherever we went when Asian girls were present, he got an enormous amount of attention. We couldn't figure it out until we saw him beside a statue of Buddha. They could have been twins.

So many names have faded from memory, but others remain. I will never forget Don Casey, Harold Terry, Roger Hurley, Art Zachary, Terry Moote, Ron Schooley, Major Kovensky, Peter Umbras, Hutchinson, Combs, Gordon Shelly, James Harris, Wally O Campo, Whitehead, Farino, Metzer, Dersky, and others who helped guide me through my year in Vietnam.

I was very fortunate to be in the Air Force and to have a job that required an air-conditioned office. I realize that 99% of the time I was in no danger. My job was to support the men of the Air Force who did go into harm's way. I was in a combat zone, but I was never in combat. I am in awe of all those who were, and wear, not a chest full of medals, but a small blue and silver badge—the Combat Infantryman's Badge. These men are heroes. Each person awarded this badge faced bullets directed at them.

One of the memories of Vietnam that many of us hold dear are the "care packages" from home. Most of us in-country received packages from our families, and they made all the difference. I couldn't wait to get a package filled with goodies. Yes, they were good, but mostly they were a physical connection with home. Whenever I got one, I shared it with my friends, and everyone shared theirs too. We all looked forward to them. I remember one that an aunt sent me, containing a prank gadget, A Bag of Laughs. A mechanical laugh, it was passed around the barracks to bring a smile to everyone, a laugh.

Probably the biggest thing we looked forward to was mail call: letters, cards, and words from home. Though the protest and anti-war sentiment were going on even before I went to Nam, there were others stateside that supported soldiers, airmen, sailors, and Marines. College students from a small school in Pennsylvania, Slippery Rock State College, had a program writing letters to *in-country* troops. A kind girl corresponded with me for quite a while. It was so welcome. It was so appreciated. I can't remember her name, and I wonder if she remembers the airman (airmen) to whom she brought a ray of light. Mail, care packages, and letters of support from strangers—all were a touch of home.

My memories of my tour in Vietnam are hazy and fading, and yet some are as clear as yesterday. Long hours of tedious work, short periods of incoming rockets, a dash for the safety of a bunker, the friendship of men who came and went, dogs, touch football, cheap beer and liquor, and cookouts.

I don't know if these thoughts of my time in Vietnam will be read, but perhaps they will give someone insight into one airman's year in a country that no longer exists, except in the minds of those who fought for its right to live free.

(Editor's note from Bob: Great story, Hugh. I consider Pleiku as my base in Vietnam, home of the 4th Infantry Division. More important-ly, our joint connection to punched cards and computers is what draws me to you. I worked for IBM for 34 years, starting in September 1968, a little over a year after I got home from Vietnam. IBM was very good to me — and still is.)

MIKE STORY

BRANCH OF MILITARY AND JOB HELD: US Army
DATES OF MILITARY SERVICE: Unit(s) served with in Vietnam: 101st Airborne Division
DATES SERVED IN VIETNAM: 1970 – 1971
HIGHEST RANK HELD: 1LT (O2)
PLACE AND YEAR OF BIRTH: Rockford, IL – 1946

WHAT THE ARMY MEANT TO MY LIFE

I am a member of the Atlanta Vietnam Veterans Business Associ-ation and "I'm Ready to Talk."

I spent three years in the US Army, and in 1970/1971 I was an Infantry Combat 1st Lieutenant Platoon Leader in the 101st Airborne Division along the DMZ in Vietnam where we had re-cently replaced the Marine Corps. I was stationed at Camp Evans and took my platoon out on search-and-destroy missions for 28 days at a time, with four days after each mission back at home base for a year. We were "Air Assault," so we were helicoptered out to our mission location, did search and destroy missions for 28 days, and then were helicoptered back to home base for four days.

Since the war was winding down, we did not see a large amount of combat, but did have five or six missions where we ran into serious ambushes or received gunfire. On my worst day, my medic was killed, and an entire squad of six soldiers were medivacked out after being seriously wounded in an ambush. That day will live with me for all my life.

Interestingly, I am still friends with the sister of my platoon medic who died that day, as I met her when I was invited to my medic's memorial service back in the States in 1971 after I was discharged.

What I really want to share for this book, though, is how much serving in the Army, and in Vietnam, meant to me in my life, how much it helped me grow, and how much it helped me be effective in life in general, especially in my career after Vietnam as a senior Product Supply Manager for Procter and Gamble for 33 years.

The Army taught me what real leadership is: the importance of knowing well the people under your command and respecting them if you wanted them to respect you. About six members of my 25-man platoon had been ordered by a judge back in the States to either join the Army or go into jail for a minor crime they had committed. They chose joining the Army.

They were good men, and some of my best soldiers in many ways. Overall, I learned so much in the Army about the importance of treating your men with respect and honor and to listen to their input. The lessons in good management and good leadership were simply invaluable.

I look at my three years in the US Army as three of the most important years of my life. I had many friends avoid the armed services, and, coming back from Vietnam, I faced a nation strongly protesting our involvement in Vietnam. Yet, I consider those three years invaluable in my life. The knowledge the Army provided me, the learning from my training and experiences, and my growth as an individual, the Army helped provide for me.

I rarely talk about my Army experiences, and almost never talk about Vietnam. This is partly a learned habit from deciding to "get on with MY life" when I returned from Vietnam, and partly because I consider my Vietnam experiences deeply personal. I also have a bias against just telling "war stories." It is too personal and not a part of "getting on with your life." Also, I deeply value my friends and the camaraderie at the AVVBA. It is a terrific organization, and I feel honored to be part of it.

So, I thank the Army, all people who I underwent training with, and all people I served with, for the irreplaceable wisdom, camaraderie, and training I received. It changed my life, it was a foundation for the success I had in both my personal and business life after the Army, and I would not trade it for anything. I also thank the AVVBA for all they have done for me.

WITHDRAWAL PHASE: 1970-1972

MAY 4, 1970: Kent State shooting. Students were protesting government policies in the Vietnam war when National Guardsmen fired on them. Four were killed and nine wounded.

JUNE 24, 1970: United States Senate votes to repeal the Gulf of Tonkin resolution.

JUNE 13, 1971: Pentagon Papers were first published by the *New York Times*. They revealed a broadened scope of the war, but had failed to tell the American people or news media about the decision. This unannounced expansion of war deepened the distrust between the American people and their government.

JUNE 22, 1971: President Nixon signs the 26th Amendment, lowering the voting age to 18.

MARCH 30, 1972: Easter Offensive launched by North Vietnamese on South Vietnam.

MAY 8, 1972: President Nixon launches Operation Linebacker I, which includes massive bombing of enemy forces in South Vietnam, resuming bombing of North Vietnam, and mining of Haiphong Harbor.

DECEMBER 18, 1972: President Nixon orders Operation Linebacker II, a massive new bombing of North Vietnam to force Hanoi back to the bargaining table and to impress the South Vietnam government with American resolve.

Source: www.vvmf.org/VietnamWar/Timeline

NATHAN CRUTCHFIELD

BRANCH OF MILITARY AND JOB HELD: Army — Armored Recon
Platoon Leader, XO
DATES OF MILITARY SERVICE: May 1968 — April 1971
UNIT SERVED WITH IN VIETNAM: 11th Armored Cavalry Regiment
DATES SERVED IN VIETNAM: December 1969 — December 1970
HIGHEST RANK HELD: First Lieutenant (O2)
PLACE AND YEAR OF BIRTH: Atlanta, GA, 1947

A DAY WITH AN ARMORED CAVALRY PLATOON

Some thoughts filtered through the mists of time.

Location: South of the Cambodian border and near Quan Loi, inside an infantry company-sized 105 battery firebase, providing additional security.

Time: Day begins at dawn

0530- 0600: Platoon begins stirring around. Quiet night. Gear stowed away.

0700: Platoon has 30 or so troopers operating six M113s (ACAV) and three Sheridans (Tanks) and is ready to roll. Number of vehicles varies depending on what's running, battle damage, and crew available. Column forms and we bid infantry goodbye, moving back to Troop Night Defensive Position several miles up highway QL13.

0800: At NDP (Night Defensive Position) — in time for breakfast, coffee. Get instructions from the Captain. Platoon is to do convoy security and escort to Quan Loi, the forward airfield and Regimental HQ.

1000: Move to designated locations to wait for convoy of trucks. Spread out along roadside and wait. Sit in sun on a hot armored vehicle and wait. Scooters with a "mama-son" carrying baskets of odds and ends rolls up. Combs, mirrors, cokes, beads, weed, etc. to sell to bored GIs. Heat feels like South Georgia in summertime.

1030-1300 (more or less): Wait for convoy. As it moves by, platoon falls in behind trucks and move through the dust until we reach Quan Loi. The red dust is like facial powder and coats everything.

1600: Move to Troop NDP. Clean up, dinner at 1730. Refuel (Top-off), do maintenance.

1800: Return to infantry firebase before dark.

Repeat.

Great news! Orders have changed! We are to begin armored reconnaissance (Recon) in force, moving through rubber tree groves, also "busting jungle" i.e. moving through brush looking for NVA or Vietcong camps, trails, supplies.

0530-0600: Day begins with breakfast from field kitchen.

0700: Briefing by captain on where we're going. Relay info to platoon sergeant and lead scout. We mount up and vehicles crank up.

0800: Column of two armored platoons with captain's and forward observer's tracks heads out. One platoon stays at NDP with HQ platoon, mortars, mess cooks, troop maintenance, and support. Move to designated area, captain orders—"turn off road" and we bash into woods.

1200: Halt for lunch—C-rations—avoid the ham and limas (beans) if possible.

1300: Rambling through trees and brush. A LOH (light observation helicopter) joins in and flies treetop level, following a trail. Pilot radios a warning, "Be very careful, trail is real hot. Looks like bunkers, Be car.... Taking Fire and going down."

Platoon bashes forward and finds smashed helicopter on side. Two crew members slightly injured but okay. Two enemy are now in a bunker. A polite request is made for them to give up. They respond, "No." A Sheridan (light tank) is rolled up to the bunker opening and fires a 152mm high explosive round into the bunker. Game over.

1600: We wait for smashed helicopter to be removed and return to NDP.

1730: Dinner, refuel, get mail, more C's for lunch, add sodas and beer to ice coolers.

1800 or so: Place flares with trip wires and claymores out for security.

Dusk: Bed down for night—cots behind vehicles, or sleep inside ACAV. Guard rotates every two hours, sitting in the cupola behind a .50 cal machine gun. Starlight scope used as it gets dark, if it works.

Dawn: Repeat.

Days vary—daily resupply as Chinook helicopters bring supplies, parts, beer and sodas, returning troopers and newbies....

While patrolling, mines might be hit, blowing road wheels off and require vehicle to be dragged back to NDP. Crew is flown back to a forward aid station or on to evacuation hospital.

My track hit a "Bouncing Betty" mine which launched by design to about nine feet up or eye level for us riding in the ACAV. It hovered for a moment, about the size of a can of veggies and dropped back down to the ground, unexploded. War can be luck of the draw.

Rocket propelled grenades (RPG) are fired in ambush and leave a small hole where they penetrate the metal. They cause severe injury and death to those inside a vehicle. Rockets and mortars fired at NDPs caused death and injury.

Occasionally, we'd come across where a major battle had occurred, with debris, helmets, bits and pieces of military gear. It was very sobering to drive by the rusted remains of a Sheridan with its turret 30 feet or so from its main body.

Each day was different, as well as the same. Boredom balanced by terrible sights; camaraderie balanced by loss.

GALE BEACHUM

BRANCH OF MILITARY AND JOB HELD: US Army, Infantry Squad Leader
DATES OF MILITARY SERVICE: November 1969 - November 1971
UNIT(S) SERVED WITH IN VIETNAM: 5th Mechanized Infantry Division and 1/22 Infantry
DATES SERVED IN VIETNAM: March 1971 - November 1971
HIGHEST RANK HELD: Sgt E5
PLACE AND YEAR OF BIRTH: McLenansboro, IL 1948

INDUCTION

I was in my third year at Southern Illinois University in Carbondale, IL. I was trying to pay my own way, went broke, and found a job. I was inducted into the Army on November 11, 1969—yes, Veterans Day—along with four others from my hometown of Mcleansboro, IL. We were all drafted. Rodney Pryor was one classmate; you'll hear more about him later.

It was at the height of the antiwar movement. I didn't want to go. I completed basic training at Fort Leonard Wood, MO and AIT in Ft Ord, CA. For some reason, I was selected with four others from my company to go to noncommissioned officer school at Fort Benning, GA. Maybe it was because I had an associate's degree and was an expert on the M14 and M16 rifles. My granddad taught me how to shoot.

Leadership and Jungle training was intense. After spending a few months with a Basic training company in Fort Polk, LA, I got my orders for Vietnam, flew to Saigon with Flying Tiger Airlines, then a C130 to Quang Tri, I Corp, 5th Infantry Division (Mechanized).

ALPHA FOUR

Assigned an armored personnel carrier painted with "Brimstone Bad MF" on the cupola. Arrived at Alpha Four, the northernmost firebase in South Vietnam. The firebase was maybe a hundred yards across. Big guns were dug in around the perimeter, surrounded by razor wire and minefields. Between the big guns were bunkers half in the ground with sandbagged roofs, where my squad holed up when we were in the perimeter. I met my squad. They were faded and battle worn.

So here is their new squad leader in new fatigues and boots: not only an FNG (F***** New Guy) but also a Shake and Bake, the term applied to those of us who got our Sergeant stripes at NCO school. My reception was chilly and disrespectful. They froze me out.

After a couple days, I got my first combat advice from Mole. He got his nickname from being a tunnel rat in II Corps. I was glad to have an experienced tunnel rat on my squad. His advice was "Don't eat apricots." Apricots were one of the fruits in C-rations.

The C-rations were vintage Korean War, dated 1951. Apricots invited incoming. C-rations had a meat can, a cracker can, fruit, four cigarettes, toilet paper, and a chocolate bar. I had a bottle of Heinz's 57 sauce I used to give some taste to the C-rations.

We should have been supplied with a toothbrush and toothpaste. Dental hygiene was non-existent. I came home with rotten teeth and spent months in a dentist chair. My wife says that she'll scrap me out when I die, with all the metal in my mouth.

Went on a RIF (Recon In Force) the next day with my platoon, and we got hit. Well damn, I thought I'd get some transition time. It went downhill from there the rest of my time. Got my CIB (Combat Infantry Badge) the first day. We never fought the VC. The NVA were our enemy and were well equipped and were good fighters.

NVA BUNKER

The platoon was patrolling far north of Alpha Four. I think we were in North Vietnam. I saw a green wire close to the trail. We cut it and followed the wire to a well concealed, well-built bunker. It was freshly built. We tossed in a frag (fragmentation grenade). It was not occupied. My squad was assigned to blow it up. We packed it with C4. Nothing but a hole was left.

As we patrolled on, we took a break, and my medic spotted a cache of mines, rockets, and weapons. We gathered them up, placed them in a clearing, and called in a Cobra gunship that lit them up. Another huge explosion. Was a successful RIF.

My medic carried an M16 and knew how to use it. My brother-in-law Patrick Dunlap was a medic in Vietnam. We've talked little about our experience. I think he saw more bad stuff than I did. He delivered a Vietnamese baby once.

He and I have a brick, located at the Peachtree Corners Georgia Veterans Memorial, along with my dad who served in the Army Air Force in Italy in WWII.

BREAK SQUELCH TWICE

As the squad leader, I had the map and compass. I had been trained, but was not prepared to navigate in dense jungle or bombed out and a defoliated DMZ. I was lost much of the time. We hated setting up ambushes. My platoon sergeant would give coordinates a few hundred yards outside the perimeter. Sometimes we'd be dropped off from our armored column. I'd do the best I could to set up in the right location, but my priority was to find a good defensive position.

I'd form a small perimeter, set out claymore mines, conceal ourselves, lay out hand grenades, flares, and shoulder fired rockets. Ambush was always at night. I didn't call in marking flares because I didn't want to give away our position.

Sometimes the perimeter would open fire close to our position. You could tell it was close with the rounds wheezing through the elephant grass. I'd radio back to cease fire. We rotated watch. Every hour the perimeter would radio and say, "If you hear me, break squelch twice." That verified we were still there.

I had one guy who frequently fell asleep. I knew he was on drugs... stupid. Once on a RIF, he started screaming, took off his clothes, and threw his M16. I choppered him out of the field and never saw him again.

We were bait on ambush, early warning, we knew that. I tried not to engage, mainly because I didn't know how large of a force I heard and did not want to expose us to unnecessary danger. I'd radio the perimeter, and they would fire artillery.

I finally got the respect of my squad. I did what I could to protect them. I'd walk point if I had no volunteers.

INSIDE THE PERIMETER

Perimeter duty was pretty safe. We'd occupy our bunker, along with the rats, snakes, huge one horned cockroaches and scorpions. We'd rotate the watch. Cannons would fire when movement was detected. We'd get in position and watch for an assault. Enemy forces were usually small, and they never breached the perimeter while I was at Alpha Four. We had frequent rocket attacks, almost daily at times. Maybe launched from north Vietnam. We were close enough to see the NVA flag with binoculars. There was an underground kitchen that served our troops, but we usually didn't go there.

The next firebase, Charlie Two, was south of us, maybe two clicks (kilometers). We frequently manned the bunkers there. Once, while under heavy rocket attack, a rocket penetrated the sandbagged roof of the kitchen and killed 21 soldiers in the mess hall.

My squad was on the perimeter during that attack. I later learned that the attack made headlines in the New York Times.

We'd spend two or three weeks in the field and return to Quang Tri for three days. We had bunks in screened barracks, hot food, and showers. While in Quang Tri, a service was held for those that were killed. Their boots were lined up in front of our formation.

WAVING BUSHES

Our platoon occupied a hill providing security for an ARVN force moving through a valley below. They were hit, and F4 jets were called in. The column got through the valley.

After we left the hilltop, we were called to return to the hill. It was dusk, and the wind was blowing hard. I was walking point with the lieutenant. The bushes were blowing back and forth, disguising an NVA patrol cresting the hilltop. We couldn't have been 20 yards apart, and we spotted each other about the same time. All hell broke loose. They fled, but as we crested the hill we saw the bodies.

We set up another small perimeter on the hill. That night we heard moaning from a tree line. The next day a chopper flew in a scout dog and NVA interpreter. The interpreter tried to get the wounded NVA soldier to surrender. Eventually he let the dog loose. The dog pointed him out, and as we advanced to his position, he opened fire. We mowed down the tree line.

The scout dog went berserk and bit his handler in the nuts. He dropped his pants, and there was a pierced hole in his scrotum from the bite. They choppered them out, and I always wondered what happened to that team.

IT WAS A BLUR

A few months in the field and the squad was choppered back to Quang Tri, then flew to Saigon in a C-130 and boarded a jet to Thailand for a three-day R and R.

Drugs, liquor, and girls. From combat to "luxury" was overwhelming. No wonder some guys went AWOL on R and R. We had a lieutenant go on R and R in Hawaii to meet his finance. He never returned. That's about all I have to say about Thailand.

PURPLE HEART

Attached is the paper my 7th grade son wrote for a class assignment. He told the story of my Purple Heart better than I can. I kept the original handwritten report. It follows:

The Diversion
By Paul Beachum, 7th grade, December 19, 1991

It was June 21, 1971, when my father was on a RIF (Recon In Force—that meant he was going out to secure an area, in this case a hilltop.)

His platoon set up in a perimeter on the top of a hill, guns pointed outward. There were about 30 to 40 men, eight armored tracks, and one tank in the platoon.

My father had his binoculars, as did the other track commanders. As he scanned the area, he came across a sandy spot where he was intrigued by some ants, busy at work. The momentary diversion soon was changed.

Out of the corner of his eye, my father saw a flash and heard a "thoomk." Simultaneously, rife fire popped around his head. He dropped down into the turret in his track thinking, "They are not going to get another chance." He yelled, "Everyone off the track!"

When everyone had cleared the track, someone yelled, "I'm hit!" The medic shouted to my father, "Beach, my bag, my medic bag." The medic had left his bag in the track. My father raced in, grabbed the medic bag, and ran out the back of the track.

As he leaped out, the second rocket hit, propelling my father into the air and sending shrapnel into his right arm and shoulder. He later received the Purple Heart for this. After the blasts, my father scrambled into a bomb crater, 30 feet in diameter and 20 feet deep, where he was pinned down by enemy fire. The medic, who had received his bag, knew where my father was. He kept shooting grenades over the hole to keep the enemy off my father.

For a moment, my father heard thousands of rounds of ammunition and grenades firing off his exploding track. Then the F4 Phantom jets flew over, spraying the area with bombs. When my father crawled out of the hole, he only had his pants, boots, dog tags, and M-16. That was all he had left from that destructive battle.

Sergeant Beachum resumes to wrap up the story: "We had been pretty beaten up and were sent back to Quang Tri to recover and get reorganized. When I jumped off the track, Rodney Prior was there, looking like he had seen a ghost. It had been reported back at headquarters that I was a KIA. He had been sent to identify my body."

GOING HOME AND CLOSING THOUGHTS

I was choppered out of the field in a Chinook and was soon on Flying Tiger Airlines headed home. The cheer and roar that erupted when we lifted off from Saigon shook the jet. Stewardess served us on the plane. First time I'd seen round eyed girls in months. It was exhilarating.

I kissed the tarmac in Seattle. I was discharged November 11, 1971. I served exactly two years. While I waited for my flight to St. Louis in Los Angeles, people stared at me and avoided me. It was later I understood why we were hated as Vietnam soldiers. As I melted back into college, I kept it to myself that I was a Vietnam veteran. Somehow, good sense prevailed, and I met and later married my wife, Nancy, and finished my bachelor's degree. I would have never made it mentally or emotionally without her.

I still have my map, an NVA helmet, a whistle, tweezers made from GI mess gear, and an NVA propaganda flyer titled 'Yankee Go Home.' I guess I need to donate them to a museum. After a few years, Nancy convinced me to buy and wear a Vietnam veteran cap. She tells the story when out for dinner someone came up to me and said thanks for your service. I hid my cap, ashamed. Later, I was wearing the cap and got a discount. I wear it all the time now.

I have hearing loss and hearing aids provided by the VA, and I visit a dermatologist every six months to burn off pre-skin cancer spots—probably a dozen or so every visit. VA said over-exposure to the sun.

Nancy says I have PTSD. I never thought I did, but she knows me better than I know myself. She labeled me The Iceman. Unfortunately, combat tends to harden you. I anguished for years about why I survived the war with so many close calls and while tens of thousands died in combat. One of my saddest moments came when Saigon fell. I cried. Guilt complex is real.

Nancy and I are still married after 51 years. I am so fortunate to have met her. We have two sons, Paul and Brandon. Paul and his wife Candace adopted two dwarf kids, one from China and one from Ukraine. In those countries, special needs kids are taken to orphanages at birth and sent on their way at 18 years old. There's no telling what would have happened to them.

So, I made some peace with the knowledge that maybe God saved me so I would meet Nancy, and Paul and Candace would adopt Lizzy and Dylan and rescue them. It gives me some closure. The war still haunts me, but I am grateful for the life God has provided for me.

DAN BENNETT

BRANCH OF SERVICE: USAF — Flight Engineer
DATES OF MILITARY SERVICE: active and reserve: 1967 to 1975
UNITS SERVED WITH IN VIETNAM: 4th Special Operations Squadron and the 18th Special Operations Squadron — Da Nang Air Base and Nha Trang Air Base South Vietnam
DATES IN VIETNAM: 1967-1968 and again 1971 to 1973
HIGHEST RANK HELD: Technical Sergeant, E-6
PLACE OF BIRTH: Atlanta, Georgia

DESTROYING AN ENTIRE COLUMN OF NORTH VIETNAMESE TANKS

On the night of February 28, 1971, I was Flight Engineer on an AC-119 Stinger gunship that notched a big kill when it struck an enemy tank convoy near Hill 31 Laos, destroying all eight Soviet

PT-76 light amphibious tanks in the convoy. We touched off 15 secondary explosions and three sustained fires.

I have to admit that we were not too happy about our mission orders that night. We were truck hunters and didn't relish flying armed escort missions. During a major operation, Operation Lam Son 719, we were assigned an armed reconnaissance for an ARVN (Army of the Republic of South Vietnam) convoy moving towards Chiphone in the Republic of Laos. Hill 31 had been the scene of extremely heavy fighting for the ARVN. If memory serves me right, the mission started out pretty boring. It was so boring that the pilot decided to break the monotony by leaving our orbit above the convoy and exploring further down the highway. That's when the mission became really interesting.

As we made a few lazy orbits above the route structure our FLIR (Forward-Looking Infrared) sensor operator picked up two hot spots moving slowly down the road. They were a few miles ahead of the allied convoy and moving in the same direction. There was no doubt that these were tanks, we just didn't know who they belonged to.

We checked in with the South Vietnamese convoy and as far as they knew there were no "friendlies" ahead of them. According to our Navigator/FLIR Operator, Major Frost, if they were enemy, it looked as if they were looking for a better location to ambush the convoy. We had to be sure, and we called for a FAC (Forward Air Controller) to identify whether they were friend of foe.

The excitement and anticipation were unbearable, and we just knew they were North Vietnamese. We wanted to take them out, and we wanted to do it now before they turned their fury on the friendly convoy. We also knew that up to this point, no gunship had destroyed any tanks in the war, or at least that is what we were told.

It seemed an eternity before the FAC arrived and all the while we waited, we could hear a nearby AC-130 Spectra crew begging to replace us. Their claims that we didn't have the firepower to tackle tanks and they did have the firepower. Their call fell upon deaf ears as we proceeded onward. The AC-130 was held off, and the AC-119K Stinger gunship was about to do what others thought it couldn't do… bust tanks!

We followed the tanks down the road. Three more large hot spots rolled out from a tree line and joined them. Before long, an additional three hot spots joined in, and they formed up at a T intersection. Up to this point, they had no indication that we were following them. I remember commenting to the crew on the radio that they might pull off the road if they detected us. Tanks have been known to back into a ditch or crater to get better muzzle elevation. And they just might be crazy enough to take a pot shot at us. The words were hardly out of my mouth when one of the tanks pulled off the road into the bushes.

By this time, the FAC had arrived and was zipping down the road at antenna top level making sure there were no Allied troops in the area. Our adrenaline soared as he screamed into our headsets. "Take them, take them, they are enemy." We had our long-awaited clearance, and the North Vietnamese tanks were about to feel our sting.

Setting our sites on the lead tank, our sensor operators acquired our target, and the pilot jockeyed our aircraft into the strike position. Our Pilot, Major Glass, called for a gun and in a heartbeat a minigun was online and control given to the pilot who immediately fired a marking burst to see if we were on target. We were, and the gunners added a 20MM Vulcan cannon to the firepower of the miniguns.

As we continued to fly in our tight target orbit, our gunners were busily replenishing the guns as we hammered the tanks with a searing mix of armor piercing and high explosive incendiaries rounds and mini ball tracers, observing a large secondary explosion on the lead tank. The enemy tank column stopped dead in its tracks.

During the entire strike, our aircraft received intense ground fire from numerous enemy small arms and antiaircraft artillery (AAA), but we continued our strike, rolling in and out of orbit in a deadly game of aerial dodge ball. We then set our sights on the trailing tank, trying to box them in and did so with a direct hit.

"Brrfrrap" and "Wham," both ends of the column were now in flames. Unfortunately for the enemy, they made no effort to scatter, and from that point on, it became a turkey shoot as we pounded them with 20 MM armor piercing rounds. One by one each tank experienced secondary explosions and all were left burning brightly in the dark Laotian night.

Having survived another night of intense antiaircraft and small arms fire, our mighty Stinger lumbered for home. As we nervously, and somewhat giddily, celebrated our success and tallied up the rounds of antiaircraft fire we received, the realization set in that this Stinger crew had just prevented the possible annihilation of a South Vietnamese convoy, by methodically destroying an entire column of North Vietnamese tanks.

RECOVERING WITH A RUNAWAY PROP

While on a mission over Pleiku Province, our "Shadow 7 crew" (AC-119 Gunship) received a call about a large group of North Vietnamese troops moving through the area. As we approached the target zone, we could see the NVA troops moving among the trees. The (FAC) forward air controller gave us clearance to fire, and we opened up. After approximately one hour of firing on the target, all ground actions slowed to a halt.

As we pulled off target, we began taking heavy fire from a

51-caliber machinegun. One of the rounds hit the right engine propeller oil line, severing the line and causing a "runaway propeller." The aircraft immediately rolled and yawed left, and we began losing altitude.

As every Shadow crew member knows, the AC-119 isn't supposed to be flyable with a runaway prop. While the pilots struggled to gain control, I (as flight engineer) began running the emergency checklist. Needless to say, for a few minutes there was a lot of scrambling on the flight deck.

After several long minutes, Major Rabinski (pilot) gained directional control of the aircraft and arrested our descent. It was through skill, professionalism, and teamwork that we beat the odds and brought the aircraft back to a successful landing and lived to fight another day.

On another mission flying out of Tan Son Nhut Air Base in an AC-119G with Vietnamese markings; I was monitoring the Vietnamese FE (flight engineer) from the jump seat. Things were quiet for two hours, and then a Forward Air Controller (FAC) radioed us asking: "Shadow 28," do you normally trail smoke off your left engine?

The Vietnamese copilot answered, "Roger that," before going back to sleep. The engine instruments showed no signs of any problems. I went back down the ladder to the gun compartment and saw a ribbon of blue-black smoke trailing as far as I could see.

I ran back to the cockpit and started trying to identify a probable cause. The Instructor Navigator confirmed that the nearest emergency airport was Phnom Phen, Cambodia. It was approximately 30 minutes flight time away. At our request, the FAC radioed our situation and intentions to our unit at Tan Son Nhut. We figured the left engine was using oil at a rate of one gallon per minute and that we still had 30 gallons of oil remaining. The oil quantity gauge acted like a DME (Distance Measuring Equipment) to "PP (Phnom Penh).

Upon landing, the left engine oil quality gauge read zero. After landing rollout, we turned the aircraft around on the runway to taxi back to the parking area, but the smoke was so thick it obstructed our view, and we could not see the taxiway. Fearing a possible fire, we shut down the engines, evacuated the aircraft, and waited for emergency vehicles. After several minutes, a wooden-wheeled fire truck arrived. It must have been a 1930's vintage fire truck. Astonishing!

After 20 minutes or so, the smoke cleared, and we decided to start the good engine and taxi off the active runway. The pilot contacted Flight Operations at Tan San Nhut and after several hours of waiting, a C-130 picked us up and returned to our base. The aircraft was left with a Vietnamese maintenance crew for repair.

February of 1974, I elected to leave the military with eight years of active duty... two tours of duty in Vietnam, was enough. I applied and was hired by a major airline as a PFE (Professional Flight Engineer). In 1982, I graduated Embry Riddle Aeronautical University with a Bachelor of Science degree and received a master's degree in aviation management / aeronautical science in 1984.

The airlines began downsizing (lay-off's). I applied and was hired by the Federal Aviation Administration (FAA) Flight Standards Field Office. I was an Aviation Safety Inspector and retired as a Branch Manager, Flight Standards Division, in 2004. My wife Catherine (retired schoolteacher) and I currently live in Cohutta, Georgia.

JON P. BIRD

BRANCH OF SERVICE: US Army – Signal Corps Officer
DATES OF MILITARY SERVICE: 1968 to 2000
UNIT SERVED WITH IN VIETNAM: 221st Signal Company
DATES IN VIETNAM: Nov 1970 to Nov 1971
HIGHEST RANK HELD: Colonel
PLACE AND YEAR OF BIRTH: Syracuse, NY– 1944

WHY I DON'T FLY IN HELICOPTERS ANY MORE

Almost anyone who was stationed in Vietnam knows that helicopters were the taxis of the air. If you wanted to go somewhere, whether it was on a mission or elsewhere, you jumped on a Huey (UH-1) and off you went. I had several interesting trips in helicopters, even if they weren't all mission related.

I had several warrant officer friends who were helicopter pilots, and one day when one of them had some free time and needed to get some "training hours" in, he asked if I would like to fly with him in an OH-6 Loach (light observation helicopter) down to Saigon and back, while he picked up some parts or documents. Who remembers?

What I do remember is eventually getting out over the Saigon River at what seemed to me to be a rather low altitude, as in hundreds but not thousands of feet. But after all, this was an observation helicopter and flying low was its mission, right? Cruising toward the city along the river at this low level made for a very special trip. At some point in time, my friend asked me, "Captain Bird, would you like to try flying this little puppy?" Well, of course I would. What possibly could go wrong?

From one document, I found this: "Loach pilots were known for their exceptional flying skills and endurance. They often operated at low altitudes, exposing themselves to enemy fire while providing critical information to ground forces." Sounds just like me — not!

Well, all I was really going to do was to operate the cyclic control stick that comes up from the floor, while my friend did everything else. Piece of cake! I was told that the OH-6 was very sensitive to pilot input, so I was to treat the cyclic stick with respect and hold it lightly. Sure enough, after a few false starts, all I had to do was to "think" about going left and I went left, or think about going right and I went right. Down below us, I could see a Vietnamese Papa-san in his small sampan, poling away up the river and minding his own business.

I think I said to my friend, "Look at that guy poling his boat up the river," pushing the cyclic stick forward as I was commenting, heading down toward the sampan. Just at that moment, Papa-san apparently decided that he had to cool off in the Saigon River, as we buzzed his sampan. My Warrant Officer friend quickly took over the OH-6's controls, with a "I think we're done flying for the day, Captain Bird," and that was the end of that adventure.

FLY THE FRIENDLY SKIES OF VIETNAM AIR

This is a much shorter story. I was visiting some of our photo detachments and had to go from Qui Nhon to Pleiku, but I was having trouble finding transportation. I think I had made the reverse trip once before in an M35 Deuce and a Half (2 & 1/2-ton truck) but didn't wish to repeat that harrowing experience. I had a Vietnamese Captain friend who said he could get me a helicopter flight there, so I said, "Sure, why not?"

Well, I was escorted to a Vietnamese UH-1 helicopter, and unlike my Saigon River experience, these pilots flew at what must've been almost maximum altitude and what seemed to be a very low speed, perhaps due to that altitude. The weather was clear, but no English was being spoken, so when I looked down on the topography and saw artillery gun emplacements, I was concerned but could do absolutely nothing until we got to Pleiku, which seemed to take forever. I think this was the only time I ever flew with the Vietnamese. I could've gone from saying "cảm ơn," (thank you) to "chiêu hồi (I surrender from the famous Chieu Hoi Program) in an instant.

FRUIT STAND, WHAT FRUIT STAND?

Our 221 Signal Company (Pictorial) provided photographic support throughout Vietnam to a variety of units, organizations, and people. One of our more interesting projects involved helping a Doctor Robinson from Ohio, who volunteered to serve the Army in Vietnam so he could study medivac procedures and adopt them to his home state.

Dr. Robinson was given the rank of Lieutenant Colonel and our informal support mission was to provide him with still photographic and motion picture support for his use once he returned home. Ohio may have been the first state to use helicopters for civilian medical evacuations from car accidents and the like.

We had several officers working on this project, and I helped a bit when I could. One evening I walked over to the Medivac helipad and jumped on an evening flight to pick up some wounded Vietnamese, somewhere west of Bien Hoa. As usual, I was armed with a 35mm camera; I think Barney Fife's bullet and pistol had been left behind.

Here's what I remember. We flew to our LZ (landing zone), where normally we would have someone on the ground guiding us and confirming we were in the right location. That didn't happen. As we landed on a dirt road, we were close to a small fruit or vegetable stand with an electric light and the light, plus the roof, just totally disappeared from the rotor wash. Oops! That's what happens when you don't have a friendly on the ground.

Once we were secured, the helicopter crew began loading the wounded Vietnamese on board the helicopter, while I was taking pictures of this event.

Next, we started to take off in the typical nose down, tail up attitude of the UH-1, and as we moved forward, the left landing skid got caught on the power lines which just happened to cross the road where we had landed and were taking off. The rest of this story is a bit hazy for me. We immediately made a hard landing, but no one was hurt, and I really didn't know where in the hell we were or how long we were going to be there.

Me? I was fully armed with a 35mm camera and extra film. Somehow, we made it back to Long Binh, but I don't remember when or how. And these examples are among the reasons why I'm no longer flying in helicopters.

MAJOR JAMES "JIM" CRAWFORD

BRANCH OF SERVICE AND JOB: US Army, Engineer Officer, Aviator
DATES OF MILITARY SERVICE: 1968 - 1973 (Active Duty), 1974 - 1982 (Reserve)
UNIT SERVED WITH IN VIETNAM: USS Hollister, Military Sealift Command Office, Manila
DATES SERVED IN VIETNAM: 1964
HIGHEST RANK HELD: Captain (O-6)
PLACE AND DATE OF BIRTH: Atlanta, Georgia, 1946

MY ARMY JOURNEY

It begins before I was drafted when I tried to enlist for the Warrant Officer candidate flight school program. I missed the 115 GT test score needed for the program by one point. The GT score required to be an Officer was 110. I point this out because after I became an Officer later, many Warrant Officers had fun reminding me of the higher requirement to become a Warrant.

Two years later, in January of 1968, I was drafted and sent to Fort Benning for basic training. I scored a 134 on the GT score the 2nd time.

My Advanced Individual Training (AIT) was at Fort Sill, OK, Sound Range Crewman, March 1968. Sound ranging was a way to detect an artillery muzzle blast or round impact by using a set of microphones in the ground evenly spaced in a straight line in the ground so the sound waves would arrive at different times on a device that looks like an EKG. I don't remember exactly how we did it, but we could give accurate coordinates of the blast and even adjust artillery fire with it.

I attended Officer Candidate School (OCS) at Fort Belvoir, VA and graduated 12/13/1968 in Class 1-69 as a 2nd Lieutenant in the Corps of Engineers. This was a very difficult course, both physically and mentally. I had a very hard-nosed football coach in high school, but those football practices were easy compared to the physical demands of the first three months of OCS. The class started with over 50, and we graduated 26. About 20 of us are still with us, and we will have our 5th reunion in the fall of 2025.

I was accepted for helicopter flight school during OCS. We were called into a classified meeting a few weeks before graduation. The Military Intelligence (MI) branch was there recruiting candidates to volunteer to serve in the MI branch instead of the Corps of Engineers. There were 26 candidates there. We were told that no one was allowed to leave until they got 20 volunteers. I told them that I was already approved for flight school and if they would honor that, I would agree to become an MI officer. They excused me from the meeting.

I was assigned as an Engineer Platoon leader, Fort Hood, TX, January—March 1969 while I waited for my flight school to start.

My helicopter flight training was in two phases. The 1st was at Fort Wolters, TX and the 2nd at Hunter Army Airfield in Savannah, GA, April 1969—January 1970. Classes 69-44 and 69-46. Hunter was the overflow from the main flight training at Fort Rucker, AL. My 1st solo flight was at Fort Wolters on May 29, 1969, in a Hellier OH-23 Helicopter. The 1st solo that really made an impression on me was my 1st solo in the HU-H Huey. I thought I was flying a bus!

My brother, Joe, a Lieutenant Colonel Infantry, Special Forces, was diagnosed with a brain tumor while I was in Flight school. My trip to Walter Reed to visit him caused me to miss training, so I was set back to class 69-46 to complete the course.

The Army gave me a compassionate reassignment to the Atlanta Army Depot, Forest Park, GA, so I could be near him and my family. I was there from January 1970 - November 1971 as a Helicopter Test Pilot. He died a month later. We found out several years later that his tumor was caused by Agent Orange. When the Vietnam Veterans Memorial ("The Wall") was built, we were wondering if his name could be put on The Wall since he died before the war ended from a condition caused by the war. The Vietnam Veterans Memorial Fund (VVMF), the founders of The Wall, came up with a way to honor him and others that died after their tour as result of conditions like Agent Orange and PTSD. It is the In Memory Program which memorializes them online forever.

I attended the Maintenance Management Course at Ft Lee, VA in 1971 while still assigned to the Atlanta Army Depot.

I got a call from my assignment officer at the Engineer Branch the day I was promoted to Captain on December 13, 1970, who told me that he needed to get me to Vietnam. He had a slot of a Huey Instructor Pilot (IP) School at Ft Rucker. Most of my flight time at the Atlanta Army Depot was in the CH-34 so I did not have the required Huey time to qualify for IP school. The war was winding down, so I did not hear from him again until late 1971 when my assignment office called to tell me he had a slot for me in the Aircraft Maintenance Officer Course (AMOC) at Fort Eustis, VA. That course was about aircraft maintenance at the unit level, and it included the Huey test pilot course.

I got to Vietnam in December 1971. With two years of depot level aircraft maintenance experience and two maintenance courses completed, I was a very qualified aircraft maintenance officer. The only problem was that there were too many of us. The war was winding down, but the officer branches were sending everybody they could to Vietnam to get their "ticket punched" so the combat tour would be in their record.

I think that there were more captains in Vietnam then than privates. I was assigned to to HHC, 765th Transportation Battalion (Trans Bn), Vung Tau, RVN. They had no slots for me, so I flew the courier run, which was their shuttle service with UH-1H that flew from Vung Tau—Long Binh—Saigon—Can Tho—Vung Tau twice a day to move personnel around, deliver parts and mail. A slot finally opened for the Commanding Officer (CO) of the Headquarters Company where I served until the unit stood down in May of 1972.

The 765th Trans Bn Commander left to go home before the stand down was complete, so he left me in charge to complete the stand down. The property book was a mess; it was missing many items, including several Hueys. It was so bad that the CO of the parent unit, the 34th Group, a full Colonel, visited me and told me to go on leave because the former CO of the 765th was being recalled to Vietnam to fix his mess.

I returned from leave in May 1972 and was assigned as the Officer in Charge (OIC) of the Night Shift Aviation Data Analysis Center, Long Binh, RVN. The job was collecting the Army's flying data from the previous day, so I prepared the aviation part of the morning briefing to the Commanding General (CG) of all forces in Vietnam.

The base was closed in Long Binh, and the 1st Aviation Brigade (Avn Bde) was moved to Saigon. I don't remember any data reports after that.

I could have gone home early then, but LTC McQueen who was the Executive Officer (XO) of the 1st Avn Bde asked me and my NCO from the data office to become the duty officer while the Brigade was moving to Saigon. I really admired LTC McQueen, so I offed to help him.

LTC McQueen asked me to be the CO of the Headquarters Company of the 1st Avn Bde in Saigon while the current CO could go home on emergency leave for 30 days. I still got a 30 day drop from my one-year tour in Vietnam in November of 1972.

During my time in Long Binh and Saigon, I served and flew with Skip Bell who became one of my best friends. He returned home after I did in 1973 and visited me for what was to be a few days. He fell in love with the girl next door. He and Carole and we were together until she passed in 2019. Skip and I were best men at each other's second marriage which worked great for both because we married the love of our lives. Judy and I are still very happy after 42 years!

My Engineer Officer Advanced Course did not start until March of 1973, so I had another short assignment at the Atlanta Army Depot, Helicopter Test Pilot from December 1972 — February 1973.

I was relieved from active duty during the Engineer Officer Advanced Course, Fort Belvoir, VA in an action called a reduction in force (RIF). The force structure was reduced for the next physical year and did not allow for attrition. I'll get to that later. I was discharged in May of 1973 at a rank of Captain.

I served in the reserves from 1974 until 1982. At first, I was with the 926 Engineer GP HQ in Montgomery, Alabama. Then I had two tours as a reserve helicopter pilot with an Assault Helicopter Company at Hunter Army Airfield in Savannah, and a Medivac platoon in Fort Sill, Oklahoma in 1978. These two tours were part of a program the Army had to supplement the understaffed active-duty units.

Remember the sudden cut in forces in 1973? In 1978, there were so few pilots in an assault helicopter company with 30 UH-1's that they could not launch more than 10 because of crew shortages. The Huey refresher course at Fort Rucker was full, so to help with this, I was in a pilot program to see if the unit IPs could give refresher training to pilots that have not logged Army time in five years. The program worked well, so I had another tour at Fort Sill, Oklahoma.

Then I was assigned as a Mobilization Designee (MOBDES) at the U.S. Army Forces Command (FORSCOM) Aviation Division, Fort McPherson, GA where I was promoted to Major. These were two week tours every year where we could be familiar with the operation of the office, in case we were needed full time in a crisis.

We were given busy work going though files manually to update the aircraft assigned by unit in the US. I asked if that data was on a computer, and they told me that the computer was only reachable by unit type, and that aircraft were assigned to all types of units. So, I was told that I could not do it. It only took a few hours to find out that the Army assigned a Line Item Number (LIN) to everything in the Army, so I turned in a request at the computer office (No PC's back then) to sort the LIN's for Army aircraft by unit. I had a computer report of aircraft assigned and authorized by unit on by boss's desk before the end of the 1st week.

I received a letter from the Army about 1983 saying that if I resigned my commission that it would assure me that if I was called up that my rank be at my permanent grade of Captain.

JOHN BUTLER

BRANCH OF MILITARY AND JOB HELD: US Army, Field Artillery Forward Observer and Battalion Fire Direction Officer
DATES OF MILITARY SERVICE: May 1968 – May 1974
UNIT(S) SERVED WITH IN VIETNAM: 11th Armored Cavalry Regiment (ACR) and 2nd Battalion, 94th Field Artillery Group
DATES SERVED IN VIETNAM: October 11, 1970 – October 10, 1971
HIGHEST RANK HELD: 1LT
PLACE AND YEAR OF BIRTH: Pittsburg, KS – 1945

THREE TOURS IN ONE

It's been said that every soldier's experience in Vietnam was unique. I don't know whether that's universally true or not, but I do know that my two assignments during my one-year tour ended up being three very different experiences.

My first assignment was a field artillery forward observer (FO) with A troop, 1st Battalion, 11th Armored Cavalry Regiment. Our company-sized unit operated on our own, only occasionally joining up for a mission with another 11th ACR troop. We moved around a lot, but operated mostly in III-Corp, the second farthest south area of South Vietnam.

We would do "sweeps" through jungle areas looking for enemy encampments or other signs of enemy activity. These operations were executed with our vehicles formed side by side 20—30 yards apart. When the jungle was dense, we often lost sight of other vehicles, so it was tricky to maintain a straight row of vehicles.

We also executed Reconnaissance in Force operations (RIFs) using various formations, usually starting in a line, then splitting off into multiple lines, depending on the terrain and other factors. This again required careful, continuous, and effective communication within our troop.

We slept in our vehicles in the jungle on whatever operation we were conducting. My "FO" team consisted of a driver, two machine gunners, and a track commander (TC) who was an E-5. Even though I outranked him, we honored his position as TC by giving him the prime sleeping spot, which was on the floor of the track. My sleeping position was on the top of a row of ammo cans against the wall of the track. When they were lined up, side by side, the flat tops of the cans were about 17" wide, so my "bed" was 17" wide with one side being the inside wall of the track. After some experimentation, I learned to sleep on my back with my arms and legs folded like a corpse.

My second assignment was a battalion fire direction officer with the 2nd of the 94th Heavy Artillery Battalion in I-Corp, the northern part of South Vietnam. My "office" was the battalion fire direction center (FDC).

This assignment ended up consisting of two completely different experiences. When I first joined the unit, my FDC was located in the middle of a firebase next to the Laotian border on the west edge of South Vietnam, just south of the DMZ. The FDC itself was in the back of a van truck located at the bottom of a man-made trench covered by timbers and protected by sandbags.

In the back of the van were three enlisted men, each with a radio, one for each firing battery. They had a "FADAC" (Field Artillery Digital Automatic Computer) located just outside of the van with its dedicated generator positioned on top of our trench. We also had three "boards" with maps with each battery at the center and the topography and grid coordinates of the coverage area for that battery. I oversaw the FDC from behind and outside of the van.

We were within range of North Vietnamese artillery, so we had to maintain constant protection from that threat. At first, I made a "cabin" above ground with a half-circle roof of metal culvert covered with sandbags. After realizing it was not very safe, another lieutenant and I created a "walk-up"—a below-ground "two-bedroom apartment" in a "Y" formation with steps for an entrance, then slits in the ground just wide enough for a cot and deep enough to crawl into with a couple of feet of space above our heads, We covered the whole thing with dirt-filled powder canisters covered with sandbags. It was cozy. Each of our bedrooms was like a grave with one open end.

When that operation ended, we moved to a surprisingly nice location between Phu Bai and Hue with plywood buildings for our quarters, a real mess hall, and a spacious air-conditioned bunker for our FDC. It was glorious! We even had latrines—plywood outhouses with several holes each. In the Army tradition, there was one for Field Grade Officers, one for Junior Officers, one for NCOs, and one for Enlisted Men. You simply can't make this stuff up!

I shared the responsibility of Battalion Fire Direction Officer with another lieutenant in two shifts, each 12 hours per day, seven days per week. For the most part, our three firing batteries executed harassment and interdiction (H & I) missions. We received targets, passed those targets to the appropriate firing battery, calculated firing data, and verified that data with each battery's fire direction center. We would then receive feedback in the form of Battle Damage Assessments (BDAs) from several sources, including Forward Air Controllers. Those reports typically included numbers of KIAs, destroyed military equipment, etc.

As far as I know, not one of us believed those reports; but we did our duty and passed that information along every morning to our Battalion Commander and his staff. Our report included a wall-mounted map with a clear acrylic cover, lights along each side, so that colored grease pencil markings on the acrylic would light up. It was very pretty, whether or not it provided any real value.

I slept in one of six spaces (rooms with half-high walls) inside of one of the plywood buildings. The building was surrounded by screened windows. My room had a cot covered by a mosquito net. I was even able to acquire an electric fan and a small refrigerator for my living quarters. The "fridge" was painted yellow and had the words "Top Secret" painted in red on it. Very classy.

MAJOR JAMES "JIM" CRAWFORD

Sometimes it's easier to seek forgiveness than ask permission.

A CHANGE OF MISSION

None of us are clear on the exact date…it was a long time ago… but we all agree it was in the spring of 1972, and the war was winding down. It was an open secret the US was withdrawing troops and equipment from Vietnam. The rumor-mill was cranking at full tilt.

Captain Jim Crawford and Captain John Wemlinger hadn't been in-country more than three months on their first tours of duty. Both were Huey pilots assigned to the 765th Transportation Battalion, an aviation maintenance outfit located at Vung Tau Army Airfield. The battalion was supposedly on its way home.

As for Crawford and Wemlinger, both were hopeful they'd get a "drop" and sent home early… but that… well, that was one of those rumors. One day everyone was going home, the next, only those with half their tour completed would catch an "early" freedom bird. No one really knew for sure.

Now, for those of you who may never have had the good fortune to visit Vung Tau, RVN during the war, it was the in-country R & R center for American troops and our allies. The rumor… there those pesky things are again… was there was also an NVA/Viet Cong R & R center as well, just north of Vung Tau, at a beach near Long Hai.

Australian OH-58 pilots believed they'd flown over it, but warned us, "Don't look like you're reaching for anything…no aggressive movements." The Aussies even said they'd waved at the beach's occupants and got waves back. So maybe there was some truth to the rumor. Maybe Vung Tau and its surrounding area was one of those places where enemies agreed to peacefully coexist.

Jim Crawford was assigned as the battalion's Headquarters & Headquarters Company Commander. Wemlinger was the Shop Platoon Leader, 330th Transportation Company, 765th Transportation Battalion.

Crawford recalls, "It was a high overcast Saturday when the 'FNG's'…that's me and Wemlinger… along with a few more well-experienced pilots were about to be pressed into doing something that each knew in their hearts was exactly the right thing to do. Yet, there was an element of risk involved."

The airfield on that day… "It was a weekend… a Saturday I think," as Jim Crawford recalls vaguely. "It was very quiet." He speculates, "Everyone who was coming to the R & R center had apparently already arrived, and it didn't seem as though anyone was heading back to their unit. The airfield ops center was empty of waiting passengers. There were no scheduled inbound or outbound flights. There were no test flights underway. There weren't even any aircraft running up on the ramp undergoing maintenance operational checks in preparation for a test flight. In fact, many of the maintenance battalion's assigned soldiers and contractors were likely downtown in the Ville …a place crawling with GI-bars…or at the R & R center's magnificent beach."

Crawford continues, "I was headed over to the Vung Tau Area command headquarters to discuss some detail of the battalion's stand down when I encountered a Special Forces (SF) captain driving alone in a jeep. He was in a predicament. He was part of a SF team training Cambodian soldiers at a base camp north of Vung Tau, near Long Hai. They were nearly out of 5.56 and 7.62 ammo, and he was sure there were VC in the area, at least two companies worth.

He told me support was supposed to come from MACV, but repeated calls for assistance had gone unanswered. Helicopter support was what he needed to get ammo from Long Hai into his team in the boonies. I jumped in the jeep with him, and we proceeded to see who might be around to help.

Pilots were the first thing we needed, and we were lucky to find Major Jim Hughes, Captain John Wemlinger, Captain Ray Thraikill, Captain Jerre Kirby, and CW2 Ron Miller.

Hughes is the one who rounded up Jerre Kirby, an experienced Cobra pilot. Kirby says, "I'd been flying the Cobra for 18 months in country. I was tired of it… stressed out… and when my unit stood down, a warrant officer in personnel suggested I might consider an assignment to the 765th Trans Battalion in Vung Tau. I asked if there was cold beer there."

He replied, "Yes, and a beach."

"I took the job and had been test flying Cobras since my arrival a few months before this mission came up out of nowhere."

CW2 Ron Miller remembers, "I got scarfed up by Thraikill who asked if I'd ever flown gunships. I told him I hadn't to any great extent, but I was checked out in Charlie and Mike (gunship) models. Thraikill response was, 'Good enough. Let's go, we've got a job to do.'"

Crawford says, "I found Wemlinger. Once we had pilots, the next obstacle was finding the birds for this mission. There were lots of them around, but most weren't flyable on an operational mission. Remember we were a maintenance outfit."

Jim Hughes proved to be the real help here. He was the Commander of the 330th Trans Company, so he knew what was on the ramp, what was flyable, and what was not. It didn't take him long to identify two gunships: a UH-1M and a Cobra. Both were ready for test flight, but Hughes, Thraikill, Kirby, and Miller, all experienced pilots, two of them instructor pilots, put their heads together and decided that this mission would be the test flights for these two birds.

Ron Miller recalls the Mike model he and Thraikill would fly had an M16 assembly consisting of four M-60 machine guns and the M-157, 7-round rocket pod assembly. The Cobra had likely just received a modification to install a redesigned tail rotor assembly, a major modification, which made its reliability for this mission questionable, but Jim Hughes and Jerre Kirby, who were to pilot it, decided to go ahead.

"As to guns, none of us can recall exactly what the Cobra had mounted, except its miniguns. None of the guns on either gunship were bore sighted, but the four pilots of the two gunships decided, in this case, bore sighting was overrated. Now, all we needed was a UH-1 to haul the ammo and anything else the SF guys at Long Hai might require.

"Without permission, Wemlinger and I commandeered the battalion commanders' aircraft, an H-model Huey. We looked for him briefly, but he wasn't around, neither was the XO, so we just took the damn thing."

With birds, pilots, ammo, etc. figured out, there remained now, just one other niggling issue: experience. The lift ship's crew really was inexperienced for this type of mission. Crawford had more flight hours than Wemlinger. He was checked out in the UH-1H, but was low time in that aircraft. He'd flown the CH-34 at a stateside maintenance depot in a previous assignment. Wemlinger was just out of flight school with fewer than 100 hours in-country, and all of that had been on admin/supply runs.

These two would be flying the UH-1H. Adding to the experience factor was the lengthening day. Everyone knew at least some of this mission might be at night. Wemlinger recalls, "Jim and I told the other guys that this was something where we lacked experience. But what the hell could we do about it. Time was running out. I told Jim I was willing to give it a go if he was. None of the six of us were going to let these SF guys and their trainees just hang there. For Jim Crawford and me, all we could do was trust our training and the little experience we had, and hope the hell we could pull this off."

EXECUTION IS EVERYTHING

Wemlinger recalls, "Jim and I flew our…or should I say, the *battalion commander's* H-model…to Long Hai, just north of Vung Tau. The approach and landing were routine. There was a slightly elevated landing pad… don't know if it was just a natural terrain feature or if it had been built up that way. My memory is that it sat above the surrounding terrain two to three feet. We shot our approach to it, shut down, and commenced loading the ammo. I can't be sure, but I think that first sortie from Long Hai to the SF base camp also included a few Cambodian troops. We were at max load, and both Jim and I knew the takeoff might be a bit of a challenge."

Crawford was at the controls and remembers, "I kept pulling pitch and pulling it some more. Wemlinger's eyes and mine were glued on the torque gauge. At 45 pounds, we weren't even light on the skids. We were close to over-torquing our ship as I eased the cyclic forward. I felt her wallow as we slipped over the edge of the landing pad, and I prayed we wouldn't strike the tail rotor on it. Still looking at the torque gauge, we were right at the redline.

"I'll be honest, I might have over-torqued her a little." I thought, "*I wouldn't be the first to over-torque an aircraft, and I wouldn't be the last. That's the kind of stuff that kept the 765th in business.* There was a low fence in front of us, only a couple of hundred feet from the landing pad. I can't tell you the relief I felt when the blades bit into the humid air and we slipped slowly, much more slowly than I was used to, through translational lift. If we cleared that damn fence by a foot, I'd be surprised."

Jim Crawford continues, "The first sortie into the SF training/base camp north of Long Hai was done in the failing light of day, but we could see the LZ clearly enough. We shot the approach directly to the ground because we were so heavily loaded. On the way in, the door gunners got a little overzealous. They wanted to 'light up' the LZ's perimeter. I told them not to fire unless either me or Wemlinger told them to."

Wemlinger remembers, "We had no idea how many friendlies were down there or where they were exactly. We'd already broken about every rule there ever was. We didn't need to add friendly fire casualties to the list. Jim gave the two door gunners the perfect incentive to obey him. He told them if they fired without our permission, we'd leave them here. We got two quick 'Yes, sirs'…no shots were fired."

Crawford picks the story back up, "We didn't shut down, and it didn't take the SF guys and their trainees but a couple of minutes to get us unloaded. Then we evacuated a couple of SF guys and a couple of their Cambodian trainees back to Long Hai where they asked us to sortie again, but our fuel was running a bit low, and dark was closing in on us. We coordinated with the gunships who claimed their fuel was OK. We shut down and as the rotors turned down and I got out of the aircraft, I realized for the first time that I hadn't buckled in. I was sitting on my damned seat belt. So much for using the checklist. We hand-pumped a 55-gallon drum of JP4 into our aircraft's fuel tank. Don't ask me where the hell the fuel came from or how long it had been in that barrel. We needed the fuel."

Wemlinger remembers, "We weren't nearly so heavy on the second sortie. I think we loaded up a few more troops and some rations; nothing compared to the load we hauled the first time. But by now it was dark. I easily picked up to a hover and slid through translational lift, the fence in front of us no longer a concern. The challenge now was finding the base camp and its LZ."

As the two slick pilots talked about this mission nearly 50 years after it happened, both recall the LZ as about the size of a "white-tire area" if the reader can recall those from their flight school days at Fort Wolters and Fort Rucker in the late 60's to early 70's.

Jim remembers, "John was flying. We were lights out, not wanting to give the VC that might be in the area any more of a target than our sound. The two of us decided that a high overhead approach was warranted; again, neither of us were concerned with power requirements on this sortie."

Wemlinger shakes his head and remembers, "Jim and I spotted the LZ because someone down there had the presence of mind to cup their hands around the lens of a flashlight and aim it up at us. Once we spotted it, I began the spiraling descent, but in one of the turns, I lost the light and couldn't find it again. But good luck was with us; Jim still had it. I relinquished the controls to him, and he completed the landing."

MISSION COMPLETE; NOW WHAT?

Back at Vung Tau Army Airfield after this was all over, we applauded our accomplishment… quietly though because the list of rules we'd broken this day was long. We hadn't over-torqued the ol' man's aircraft… or at least so little that neither of us decided it required a write up in the aircraft's logbook. Crawford and Wemlinger got a small, humbling taste of what "operational pilots" did every day.

We accentuated the positive as the song says, but in hindsight, there was plenty of negative that could have played out. The gunships' miniguns worked flawlessly as their pilots tested them as part of the "maintenance test flight."

Ron Miller says, "We fired into a tree line identified by the SF guys as 'suspicious.' In those days towards the war's end, the rules of engagement had changed. There were strict controls on "free-fire zones." A crew had to get permission to fire, and who knows how the hell long that might have taken on a sleepy Saturday in a war zone where the war was about over. So, we kinda' blew off the permission-thing." He chuckles as he adds, "It was just a test flight."

Jim Crawford and John Wemlinger both recall how close they'd come to the red line on the torque gauge. Both speculate they might have even slightly nudged the needle past it on that first sortie.

Jim says, "We could have struck the tail rotor as we wallowed coming off that elevated landing pad or hooked a skid on that damn fence on takeoff during that first sortie. Either of these would have caused a crash. There could have been casualties. How would that have been explained? I heard the battalion commander, once he got back to Vung Tau and caught wind of what had happened, royally chewed out Jim Hughes."

John Wemlinger remembers Hughes, his company commander. "Jim Hughes never said anything to me about getting his ass chewed. He seemed to me to be the kind of guy, a Mustang, if memory serves me correctly, who'd had his ass-chewed on more than one occasion before, so the battalion commander wasn't getting any fresh meat. More to the point, ass chewing or not, I think Hughes would do the very same thing again if the need would have arisen. He was a soldier's soldier."

Jim Crawford agrees, "Who knows if we saved lives that day? But we sure pitched in to make it more difficult for the bad guys hanging around Long Hai!"

Jim Hughes and Ray Thraikill have gone to their great reward. Crawford, Wemlinger, Miller, and Kirby are spread across the States, but through the internet have been able to communicate and reconstruct that day's events.

Wemlinger, who retired a Colonel after 27 years of service, concludes, "I don't think there's any doubt among those of us still alive… we have no regrets… like Jim Hughes, we'd do it all over again and never hesitate."

(Editor's note from Bob: This is one of the longest stories in this book, but as I read it, there was nothing I could cut out. With my limit of 2,000 words for a story—I broke my own rule. What are they going to do to me, a simple Infantry officer? Send me to Vietnam?)

JOHN BUTLER

MAD MINUTE WITH THE 11TH ACR

Nearly all the time I served as a Forward Observer (FO) with A Troop, 1st Squadron, 11th Armored Cavalry Regiment (ACR), we operated alone, with no other unit near us. We would move to a new spot on the map, set up our Night Defensive Position (NDP) and operate from that location for a few days before moving to another spot.

Our troop was a company-sized unit with three platoons, each consisting of one M551 Sheridan and five M113 Armored Cavalry Assault Vehicles (ACAVs). The Sheridan's were light, aluminum hulled tanks, each with a 6" main gun, a 7.62mm machine gun coaxial to the main gun plus a 50-caliber machine gun mounted on the turret. Each of the ACAVs had one 50-caliber machine gun mounted forward center, and two M60 machine guns mounted on the sides.

Our HQ/Support platoon had two M113 ACAVs, one M577 command vehicle, one M106 (4.2") mortar track and two M125 (81mm) mortar tracks. The mortar tracks typically set up in the middle of our NDP, surrounding the command vehicle.

Our troop commander assigned me to function like a platoon leader for our "mortar platoon" even though there no such thing in our official Table of Organization and Equipment (TO&E). This made sense for us since I was already coordinating artillery support as the troop's forward observer.

My Field Artillery training included setting up, shooting, and adjusting Night Defensive Concentrations (NDCs) outside of our perimeter each time we set up a new NDP. This involved establishing grid coordinates for a few (typically three or four) spots where an enemy would most likely approach our position for an attack.

When I first joined my troop, I established my NDCs like I

was "supposed to do" two or three times before I finally came to my senses and realized that if we were actually attacked, our own firepower would be so overwhelming we would not need to add artillery support as well. The "mad minutes" we conducted every night confirmed this fact unmistakably.

Within a few weeks after joining my troop, I asked my troop commander if he would mind if I stopped executing the NDC process. He not only agreed, but he seemed to be relieved that I would stop test-firing NDCs around our perimeter every time we moved to a new NDP.

Each night, all the troop's ACAVs (including my FO track) set up facing outward, forming our circular perimeter, with the three Sheridan's positioned between ACAVs in strategic locations most vulnerable to attack. The main gun of each Sheridan was always loaded with an anti-personnel "canister" round with about 10,000 flechettes (small metal darts about 1.5" long).

None of us really feared being overrun, but just to be extra-cautious, we added trip flares and claymore mines about 20 — 30 yards beyond the perimeter.

Every night, we would conduct two or three "mad minutes" during which time we would fire all of our machine guns into the jungle surrounding us. I strongly suspect that the primary reason this "tradition" was established and maintained had much more to do with keeping us awake than deterring the enemy from a ground attack. Regardless of the reason, the effect was quite spectacular!

I had a small battery-operated reel-to-reel tape recorder with 3" tapes that I used to send home voice recordings from time to time. I recorded one of our mad minutes for posterity. Of course, the recording didn't do justice to the real thing, but it was fun to record and replay it from time to time.

CLARENCE "CLYDE" ROMERO JR

BRANCH OF MILITARY AND JOB HELD: US Army Scout Helicopter pilot, 1969-2005

DATES OF MILITARY SERVICE: 1969 - 2005

UNIT(S) SERVED WITH IN VIETNAM: C troop 2/17 Cav, 101st ABN Div AMBL

DATES SERVED IN VIETNAM: April 1970 to April 1971

HIGHEST RANK HELD: Colonel (O6)

PLACE AND DATE OF BIRTH: Bronx, NY 1950

SCOUT PILOT AND A NIGHT FAC MISSION OVER THE HO CHI MINH TRAIL

I was a scout pilot with C Troop 2/17 Cav flying out of Hue Phu Bai. We normally worked with our USAF FACs (Forward Air Controllers) across the runway who flew the O-2 (Cessna's Skymaster with twin tail-booms and two engines in pusher-puller configuration). I got to know them very well.

One of the pilots asked me if I wanted to go on a night FAC mission with him. They flew one month day and one month night.

Since we did not fly at night, I said sure. It was December of 1970. We took off in the dark and flew north for about an hour and a half. We started an orbit and contacted the C-130 gunship who was above us. I could not see him since I was in the back seat of the O-2 and there were two pilots up front running the operation.

All of a sudden, the fight was on!

Our aircraft was spotting trucks with a starlight scope and relaying the positions to the C-130 Specter Gunship. If you have never seen a C-130 Gunship in action, it's something!

The C-130 gunship started firing their weapons at the trucks on the trail.

What I didn't know was that there were F-4 fighters called Night Owls that were holding above the C-130 and they would roll in and drop their bombs on the flashes from the C-130 Gunship, man this was something to see!

The F-4's were totally blacked out, no lights whatsoever, diving at the ground, dropping bombs. What a show… but that's not the end of it.

The North Vietnamese anti-aircraft gunners were shooting as well, huge 37mm and 57mm guns up at us. But they could see us but man are they big shells.

This goes on for about one hour, then we fly back home and land.

I later found out I was over one of the most heavily defended areas of the Ho Chi Minh Trail. We were attacking the Ban Kari Pass! This place was a choke point and was heavily bombed during the daylight by F-4's and many were shot down in this area as well.

After that night FAC mission, I decided not to go out with them anymore.

But I was glad I went; I was also very young!

ROBERT E. "BOB" DEBUSK

BRANCH OF MILITARY AND JOB HELD: US Army Aviator, Helicopter Pilot
UNIT(S) SERVED WITH IN VIETNAM: 123rd Aviation Battalion
DATES SERVED IN VIETNAM: Jan – Dec 1970
HIGHEST RANK HELD: CW2

FLYING INTO THE FOG

"Slicks, I just got a call from operations that there is an LRRP patrol

near us requesting an emergency extraction. They're in the mountains in the fog with an injured team member slowing them down. They also believe "Charlie" is on to them and close by." That call came from our team lead. As the most experienced Huey Aircraft Commander flying that day, I took the mission and headed for a nearby secure airfield to unload the troops I was carrying to make room for the patrol.

I arrived in Vietnam on January 1, 1970, and was assigned as an Army combat helicopter pilot with the 123rd Aviation Battalion headquartered in Chu Lai, I Corps. Upon arrival at the battalion, I volunteered to fly the UH-1 ("Huey") aircraft for Company B (Warlords), an aero-scout reconnaissance company, an emerging concept in Army warfare. Our mission was to conduct daily patrols in the mountains and coastal plains surrounding the large Marine airbase at Chu Lai for signs of enemy movement. Our standing orders were simple: find the enemy and engage him immediately.

Our aerial patrols were typically a fire team consisting of a Loach (a small scout helicopter flying at treetop level), two Cobra gunships, and one or two Hueys ("Slicks") carrying Army infantrymen assigned to the unit. As a slick pilot, I was called upon to conduct several different missions. Our primary purpose was to provide security and rescue should one of our team members be forced down into enemy territory. We would also place infantry on the ground to gather intelligence in areas showing signs of recent enemy activity. But, because of the Huey's versatility, we were often called upon to conduct medevac missions, resupply troops, combat insertions, and insert or extract LRRP patrols.

On the morning of November 1, 1970, we had just started our patrol when we received the call of a six-man LRRP team (Long-Range Reconnaissance Patrol) in trouble. We were late getting started that morning because of the monsoon weather, which would be a major factor in finding and extracting the patrol who were on a mountain top enshrouded in clouds. While I was unloading my infantry passengers, the lead gunship found the patrol by homing in on their field radio using one of his navigational aids and reported that the landing zone was a rock outcropping and would be tight.

The gunship and I landed in an open field at the foot of the mountain and went over our extraction plan. Since we would be flying up the side of a mountain in the fog, we would have to travel at treetop level to keep the ground in sight; travel at a very slow speed to avoid hitting any terrain; fly in a close formation in order not to lose sight of one another to avoid a mid-air collision; and, I would also use my navigation aid to home in on the patrol's field radio to anticipate the direction we would be traveling. Flying a large and loud Huey helicopter low and slow in enemy territory went against all my training and experience. However, being nestled in nice and close to a gunship escort was very reassuring.

Our navigational aids took us directly to the patrol, and they appeared suddenly out of the fog, waving frantically. The landing zone was very tight. Guided by my door gunner to keep our tail rotor out of the trees, I hovered down to the side of the rock outcropping and slid in sideways to reach the patrol. The outcropping was uneven, and I could not set the aircraft down, so again, with directional assistance from the door gunner, I rested our right skid on the outcropping to add some stability for the patrol loading onto the aircraft. Big smiles and slaps on the back of my helmet let me know they were glad to see us.

Even though I was hovering the aircraft throughout the extraction process, I could see that beautiful Cobra gunship hovering in front of me just daring anyone in the vicinity to be foolish enough to fire on us—and they didn't! After the extraction was completed, we joined up in close formation with the gunship flying low level down the mountain, being careful to keep our tail rotors out of the trees. After returning the LRRP patrol to their unit, we picked up our infantry and proceeded on our patrol. All of this took place in less than an hour.

This mission serves to exemplify the dedication to "leaving no man behind" exhibited by U.S. Army helicopter crews during the "Helicopter War" in Vietnam.

JOHN BUTLER

CARE PACKAGES: THE GOOD, THE BAD, AND THE UGLY

Let me start by saying that it was ALWAYS great to receive a letter or a packager from home when I was in Vietnam! Getting a tangible reminder with something (anything) from home was like being a kid opening a present on Christmas morning!

But I suspect each of us remembers opening that present from under the tree with a pair of ugly socks in it—especially when we were expecting a toy. So, what kind of care package could possibly be disappointing?

First of all, it actually was NOT the condition of the package, even though what was delivered often only vaguely resembled what was sent. Nor was it the always the condition of the contents, even though it wasn't unusual to receive a package with nearly unrecognizable contents. Sometimes the nearly unrecognizable contents ended up becoming the best surprises!

However, I'll never forget the care package I received that contained a doll. That's right, I didn't stutter. It was a hand-painted wooden doll of a guy with flexible arms and legs designed to "dance" on a paddle that you sit on.

So here I am in the jungle working, fighting, living, and sleeping with four other soldiers in an Armored Cavalry Assault Vehicle with very tight quarters, almost no personal storage space, no privacy at all … and a brand spanking new … doll.

I won't say who sent it to me, but it was NOT intended as a joke so I couldn't just throw it away! I had to hide it at the bottom of the ammo can I used for personal items. And that's where it stayed, trust me!

On the other hand, I also received some truly wonderful items. My grandma Butler ("G'ma" she called herself) sent me the most positive, encouraging letters I've ever received. I kept them and re-read them multiple times. She seemed to know what I needed to hear. She was inspired … probably in large part because five of her sons served overseas in WW II.

And I received food items. My favorite dessert has always been (and still is) a rather dense chocolate cake with cinnamon in it that my mom used to make. It was sort of a "crownie" (a cross between cake and a brownie). My wife Suzie makes it occasionally, but she has to limit how often she makes it because I have a really hard time not eating it non-stop until it's all gone!

There was really no way to properly pack and transport something like Mom's "crownie" from the States to the jungles of Vietnam. She usually packed a slab of it in Saran Wrap, then in aluminum foil, then in a thin cardboard box wrapped in brown paper.

Sounds reasonable until you subject it to 100+ degree temperatures with 95+ percent humidity, crammed inside of mail bags with a hundred pounds of other packages for a week or two. What arrived would be a thin (smashed) cardboard box with all the contents blended together into a blob. It still tasted wonderful, even if the "non-crownie" ingredients had to be spit out regularly.

And no, I did not share this particular culinary experience with my "roomies" … now that I think about it, I don't actually recall anyone asking for a handful of it anyway!

What happened to the doll, you might ask? I could never bring myself to throw it away. It's still buried ... in a trunk in my attic with other items from Vietnam I never play with!

CLIFF PENROSE

BRANCH OF SERVICE: U.S. Army, Armor. Dual rated Army Aviator; Active Duty and National Guard: 1967 – 1978
UNITS IN VIETNAM: 335th Transportation, Aircraft Direct Support; Division Artillery Aviation, Americal Division
YEARS IN VIETNAM: May 1970 – May 1971
HIGHEST RANK: Captain (O3)
BORN: Henderson, NV – 1943

FIRE BASE MARY ANN: THE WORST SINGLE FIREBASE TRAGEDY OF THE WAR

I joined Division Artillery (DivArty) Aviation, Americal Division, Chu Lai, Vietnam in July 1970. I was the Executive Officer/ Operations Officer. We were the largest Aviation Section in the U.S. Army. We had 18 pilots, 15 Aircraft (OH-6A), and 40 enlisted personnel. Our missions consisted of: Visual Reconnaissance (VR), Artillery Fire Support, Special Missions with Special Forces, and various other assignments.

My introduction to the unit and its missions was quick and rude, as the Americal's 196th Light Infantry Brigade was moving back into Kham Duc! This firebase was close to the border with Laos. It had been overrun a year earlier, and the terrain and the strength of the enemy had not changed. It was a bad place. The NVA seemed to control the mountains and the ridges around the base camp. Kham Duc is another story.

During the latter half of 1970, The Americal and the 196th had been developing firebases in the western part of Quang Tin Province. The farthest west was FB (Firebase) Judy. Further east towards Chu Lai was FB Mary Ann, and five miles further east, was FB Mildred. Mildred was the only one that could be resupplied by road. This was not even the greatest route. It came from Tam Ky, which was about 15 miles to the east.

Mary Ann and Judy were even at the far reaches of aircraft support from Chu Lai, so air support could only stay on station for a short time before needing to re-fuel. All the above being said is to set the stage for the worst single Firebase tragedy of the Vietnam war.

Firebase Mary Ann was established approximately 30 miles east of Chu Lai in a mountainous area. It was strategically located near branches of the Ho Chi Minh Trail. It was manned by the 1/46th Infantry. The Battalion TOC (Tactical Operations Center) was located on this Firebase. The Battalion Commander was Lt. Col. William Doyle. No major battles were fought during the latter part of 1970, but units of the 196th Infantry Brigade had been in constant contact for long periods of time.

Some accounts of the Mary Ann tragedy say that activity around the firebase was low and had been for a while. I do know that Mary Ann had not been subject to attacks by rockets or mortars, nor had there been any probes of the perimeter.

In February 1971, there were signs pointing to a possible NVA buildup as two different companies found weapons stored in different locations. Also, one of those units had been in steady contact for two weeks.

I do know that we had been flying numerous missions in the area, and that the Infantry units had been in many contact situations during the latter part of 1970 and early 1971.

In March 1971, we had been instructed to increase our presence in that AO with aircraft performing Visual Recon and aircraft prepared for Artillery support. I assigned myself to fly these missions periodically. Most of the guys wanted these missions because of the possibility of action.

On March 27, 1971, I was flying the area close to Mary Ann when a transmission chip warning light came on. We headed immediately to Mary Ann to await a maintenance crew to take care of our aircraft. We were on the ground for a few hours before the crew could get to us, and a couple more as they fixed the problem. The maintenance crew came out in a Huey. My aircraft was of course, an OH-6, Loach. We needed to get back that evening as I was responsible for approving the assignments for the next day.

While we were there, I was able to look at parts of the perimeter defense at the request of the Company Commander, CPT Rich Knight. Company C had been on patrol and had just taken over security of the Firebase duty from one of the other companies. What I noticed was some areas that needed attention, and I gave those ideas to CPT Knight.

It should be noted that plans were in place to move the Battalion TOC to Firebase Mildred and to close Mary Ann. Some of the artillery had already been moved. The attitude of the officers and the troops seemed to be good as they were leaving this firebase soon and moving to Firebase Mildred.

This was not a good situation as the firebase had not been attacked in a long time and complacency was present.

I ate dinner there and CPT Knight asked me to stay the night. I could not, due to the operational responsibilities I had for my unit. Also, the maintenance aircraft was to follow me back to Chu Lai, just in case something went wrong.

We got back to Chu Lai around 2100 hours. I finally got to bed around 2300 hours. I was roused out at around 0300 hours and informed that Mary Ann was being attacked. We were not the immediate response aviation unit, so we waited a few hours for our mission's requests.

We were instructed to have three or four aircraft in the area to perform VR and other tasks as assigned by the units on site. When we got out there, we saw the mess! 33 killed and 83 wounded. It was the single most causalities of any firebase fight during the Vietnam War.

Unfortunately, CPT Knight was one of the KIA. A few of the officers and senior NCOs were among the KIA.

March 27th was my wedding anniversary. I will never forget Firebase Mary Ann.

LINDA PENROSE
ARMY WIFE OF CLIFF PENROSE

WHEN CLIFF PENROSE WAS IN VIETNAM (MAY 31, 1970 TO MAY 31, 1971)

During the month before Cliff left for Vietnam, we moved to Wichita, KS, my hometown. I would be close to my parents, friends, and familiar territory. Also close was McConnell Air Force Base. Being close to my parents was convenient as our two boys were five and three. Grandparents were great.

We were able to find an apartment that was only as few miles from my parents.

I was able to enroll at Wichita State University and work on finishing my degree. My college time was interrupted when Cliff started on active duty, and we moved out of Wichita. It also helped distract me from thinking about what was happening in Vietnam. I was also able to get involved with a group of military spouses whose husbands were also deployed. Mostly Air Force wives and great people and good support.

My year without Cliff got interesting fast. My first two letters to Cliff described a minor accident—USAA resolved it—and hail damage, which USAA also handled. Also, a few months later, a tornado touched down very near the apartment. It was a hectic start to the year. I'm sure Cliff was concerned about how I handled these challenges.

We did get to meet in Hawaii and travel to the island of Kauai, where we stayed at the Coco Palms Hotel. That is where they filmed the Elvis Presley movie, 'Blue Hawaii'. I was embarrassed because we immediately jumped into bed for some rest! Ha!

Well, when we went to get something to eat and came back to the room, they had made the bed and tidied up the room! Another funny thing happened at R&R: I sat next to a lady who was also meeting her husband coming from Vietnam. We arrived two days earlier than our husbands. As it turns out, Cliff had sat next to this Air Force pilot on the way to Vietnam, and on the way to Hawaii. Well, the lady had laid out in the sun too long, and she was beet red! They had a rough time during the short time in Hawaii. The 'R&R' was too short.

I went back to Wichita and continued to live as normally as possible. Friends and family helped. The life in the military was not easy, but we were a close knit and patriotic group. After Cliff returned home, he was assigned to the Armor Officers Advanced Course at Ft. Knox, KY. After that assignment, we went to Germany where he was a Tank Company Commander. It seems like I never saw him during the time he had that command.

In 1973, we made the decision to leave the Army and make a go at civilian life. We did and still do miss our time and all our friends in the military. We will always respect and support our military families.

BARRY PENCEK

BRANCH OF MILITARY AND JOB HELD: USMC, Cobra pilot
DATES OF MILITARY SERVICE: June 1966 – August 1996
UNIT(S) SERVED WITH IN VIETNAM: HML-367
DATES SERVED IN VIETNAM: May 1970 – Jan 1971
HIGHEST RANK HELD: Colonel (O6)
PLACE AND YEAR OF BIRTH: Scranton, PA – 1946

MIRACLES

I flew Marine Cobras with HML-367, out of Marble Mountain Air Facility just east of Da Nang. It was zero dark thirty on an all-too-short night in September 1970 when the duty driver entered the hooch and shined his flashlight in my face and said to get ready for a 0500 brief in the ready room. A dozen groggy pilots listened as our CO said that we would be going to Kontum in II Corps to support an Army operation. It was a Mission 72, which meant we would be crossing the border into Laos so we would be sterile — no dog tags or wallets to identify us. No other details were provided, but he said we might be gone for several days.

On 7 September 1970, we launched for Kontum, along with six Marine CH-53 transport helicopters, call sign Dimmer. After landing, we were trucked to FOB 2, the home of B-Company, 5th Special Forces. They were part of the Studies and Observations Group (SOG), a covert military unit that specialized in clandestine cross-border operations. The briefer said we would be inserting a Hatchet Force of sixteen Green Berets and one-hundred-twenty Montagnard mercenaries, led by Captain Gene McCarley. It was called Operation Tailwind and would be the largest and deepest raid in SOG history. Tailwind's mission was to disrupt and distract the enemy in support of a larger CIA operation that originated in the White House.

It was monsoon season, and we were told we would load the troops at FOB 2 and fly to Dak To, a staging area forty klicks north, to await a weather report. We spent the night at FOB 2 and launched for Dak To early the next morning in marginal weather.

Hours passed while we waited for the weather in Laos to improve. Our twelve Marine helicopters, plus six Army Hueys, call-sign Bikini, and six Cobras, call-sign Pink Panther, made for a tempting target sitting along the runway.

By mid-afternoon the local bad guys had organized a welcoming party. I was catching some rays on the roof of the bunker when the first mortar round hit. I grabbed my camera and started shooting. After some great action photos of the backside of my lens cover, I opted for the relative security of the "O Club." Once there, I got a shot of Panther pilot Pete Gotch scrambling along the ditch after his Cobra took a hit.

Several other aircraft were damaged during the attack, and the mission was aborted. We limped back to Kontum, repaired some battle damage, and returned to Marble Mountain.

Poor weather kept us on the ground for a few days. On 11 September 1970, we once again loaded up the troops and headed for Dak To. This time we had four Cobras and five CH-53s. The troops were loaded into four of the 53s and a fifth was designated as the SAR (search and rescue) aircraft outfitted with an eighty-foot extract ladder, made of steel cables and six-foot aluminum rungs, rolled up on the rear ramp.

This time we didn't overstay our welcome and the gaggle got airborne and headed north before turning west into Laos. Low monsoon ceilings and intermittent rain kept us close to the ground, so we received frequent welcoming ground fire from the jungle below.

A small pathfinder team had landed fifteen minutes ahead of us and reported the zone was "quiet as a church." It looked like an easy-peasy milk run. The LZ was only large enough for one CH-53 at a time, so we would have to make four trips in. As we approached the insert area with the first transport, "quiet as a church" erupted into a Morman tabernacle choir of gunfire. All the aircraft were getting hosed. I was flying with Bob Grove and when our radios were shot out, we had to yell back and forth to communicate. The 53s quickly sequenced in for offload. It was a small miracle that no one was killed or wounded.

McCarley had told his medic, Sergeant Mike Rose, to bring five to ten times his normal load of medical supplies. It wasn't long before Rose was very busy. Everywhere they turned, the SOG warriors ran into the NVA. Casualties mounted with every fire fight. By the end of the day, Rose had one KIA and multiple walking wounded.

Playing a game of cat-and-mouse with the NVA required fair enough weather for air support and mobility to stay ahead of the enemy. The weather, though marginal, was holding but the casualties slowed the Hatchet Force. It was becoming a challenge for McCarley to keep the SOG mouse ahead of the NVA cat.

By day two, Rose had a second KIA and was hauling two litter cases among his patients. Reluctantly, he abandoned the KIAs. McCarley's men were already running low on ammunition and the Army decided to try a medevac and resupply. Unfortunately, the lead Cobra took heavy fire and crashed and the medevac turned into an aircrew rescue. It was again a miracle that no one was killed.

Day three arrived and Mike Rose, who had been wounded the day before, had two dozen wounded to tend to, including two litter patients. All McCarley's SOG fighters were running low on food, ammo, and sleep. As casualties mounted and their ammo dwindled, a new plan was hatched to relieve the Hatchet Force.

We briefed for a medevac mission and left Kontum under cloudy skies. The lead CH-53, piloted by Bill Beardall, had two SOG medics onboard with trauma supplies and extra IV lines. It was a tight LZ, but Beardall decided to give it a try. Vertically descending between some trees, he started taking heavy fire. Beardall got within five feet of the ground and Rose began handing up the first of the litter patients to the SF medics on the helicopter when the rotor struck a tree, causing severe vibration. Before the exchange could be completed, Beardall aborted and began climbing. More heavy fire took out most of the hydraulic lines and Beardall got out a Mayday call as he headed for an open space two klicks from the LZ for a semi-controlled landing.

There was no room for the SAR bird to land so they hovered nearby and rolled out the extract ladder. Beardall and his crew and passengers hooked on and were airborne in minutes. Another failed mission and another lost aircraft, but miraculously, no one was killed or wounded. How many more of these miracles could we hope for?

That night we briefed for another medevac/resupply attempt the following day. The briefer said the NVA had been bringing in truckloads of reinforcements, many of them towing 37mm anti-aircraft guns. We were told to expect high casualties. I wrote a "to be opened in case of my death" letter to my wife and mailed it to my brother.

On day four, McCarley had his troops on the move before sunrise, and they came upon what appeared to be an NVA headquarters. After a brief skirmish against rear-echelon troops, they entered the complex. There were multiple bunkers and hooches, including sleeping quarters, a mess hall, stores of truck parts, rice, ammunition, and weapons. McCarley got a call to come to one bunker. When he entered, he found a room with map-covered walls and footlockers of documents, logs, and other records. McCarley ordered his men to fill their rucksacks with as much of the stuff as they could carry.

Realizing the Americans had probably grabbed some valuable intel, the NVA had no intentions of letting then get out of Laos alive and attacked with everything they had. It quickly turned into a nightmare scenario for the Hatchet Force. The weather was on its ass, precluding air support, and McCarley's men were rationing ammo as they found themselves surrounded by hundreds of NVA troops.

Another small miracle was needed and appeared when McCarley's radio came to life. Somehow, Air Force A-1 pilot Lt. Tom Stump, and his wingman Lt. Chris Tateisha, dodging anti-aircraft fire, heavy rain, and low ceilings, were able to locate the beleaguered Hatchet Force. When Stump checked in with the Hatchet Force, McCarley said, "We're getting our ass kicked."

Stump commenced a withering attack that allowed McCarley to break contact and escape. When all their ordinance was expended, Stump turned toward Da Nang. As he left, he called the radio relay C-130 high overhead, "If you don't get them out now, they're not getting out."

We had a quick mission re-brief in the morning, and they told us of the dire straits the grunts were in. The mission was changed to an extract. The briefer said, if we could get through the weather, and if we used teargas for the extract, we might have a chance of getting them out. We were issued gas masks just before we departed to meet up with six A-1 Skyraiders out of Thailand carrying CBU-30 teargas dispensers.

The pucker factor was high as we crossed the border. We needed another miracle to pull this off. And then it happened. The weather broke. Now, we had fixed-wing assets loaded for bear at our disposal. The air support was in concentric circles. The fast movers were furthest out, suppressing the 37mm guns while we stayed in a tight wagon wheel around the Hatchet Force. A-1s hammered the space in between. The hills were alive with the sound of exploding ordnance, the flash of tracers, white plumes of rockets, and adrenaline.

Only two of the six gas birds made it out of Thailand, but they dropped their tear gas, and we brought the first transport to the LZ. Fire was heavy but all the wounded, the intel documents, and a bunch of Montagnards filled the bird. A minute later, the second CH-53 landed amid a storm of green tracers and loaded up.

Persky got to about four-thousand feet when the second engine quit. "Mayday, mayday, mayday. Going down," he called on the radio. He was hoping, in the pregnant pause that followed, for someone to come on the radio and tell him there was a possible landing spot off to the right or left, but he was greeted by silence and a triple-canopy jungle coming up to meet him.

Fortunately, as he crossed a ridgeline, Persky saw a river at the bottom of a steep gorge with a small sandy area, not quite big enough for a CH-53, off to the left. The aircraft hit hard, split in two and rolled over. One Yard died and most everyone else was injured. Fuel was streaming from the aircraft but, miraculously, there was no fire.

We called in the SAR bird. With no place to land, pilot Jack Tucker placed one main gear on a rock in the river and lowered the ramp to the bank. The dazed, injured, and wounded men from the crashed bird limped aboard and we all headed for Dak To.

Back at Dak To, we inspected the battle damage to our bird. We had taken a hit through the tail rotor driveshaft, but fortunately it didn't fail. Another hit gave me pause. The Cobra had a small piece of armor on each side of the engine to stop small arms, and we had taken a hit right on the corner of that armor. A half inch lower or to the right would have downed the aircraft. I was glad Tailwind was over, and we wouldn't be seeing that part of Laos again.

The story of Operation Tailwind was soon buried, collecting dust somewhere in a classified archive, never to be spoken of again. Or so we thought.

In 1998, twenty-eight years after the top-secret mission ended, CNN produced a documentary called *Valley of Death* in which they accused the participants in Operation Tailwind of war crimes, specifically using nerve gas to kill women, children, and American defectors.

The broadcast had a hollow ring to it, and almost immediately veterans groups, and even some main-stream media, began to raise the BS flag. Soon, a media firestorm erupted, and CNN itself became the focus of an investigation into its questionable journalistic practices. Within weeks, after conducting their own investigation, CNN apologized and retracted the story.

There were several positive outcomes resulting from the CNN debacle. All military records pertaining to the mission were declassified, which allowed for some long overdue recognition. In April 2001, SOG was awarded the Presidential Unit Citation. In October 2017, Sergeant Mike Rose, the SF medic on the mission, was awarded the Medal of Honor at a ceremony at the White House. At the ceremony, someone asked me if I was a very religious person, and I said, "No." But, if you asked me if I believe in miracles, I'd say, "Not only yes, but hell yes!"

Semper Fi,

Barry Pencek

HAM HENSON

BRANCH OF MILITARY AND JOB HELD: US Army, Infantry Platoon
Leader and Company XO
DATES OF MILITARY SERVICE: 1969 - 1978
UNIT(S) SERVED WITH IN VIETNAM: C/1/502 Infantry, 101st Airborne
DATES SERVED IN VIETNAM: May 1971 - May 1972
HIGHEST RANK HELD: Captain (O3)
PLACE AND YEAR OF BIRTH: Savannah, GA 1947

RAT PATROL

This is an amusing account of what happened when my unit, C/1/502 Infantry, took over an ARVN firebase on the DMZ in 1971. We were assigned to relieve the ARVN unit so they could begin an offensive operation.

Upon arriving, we found the transition easy. All we had to do was assume the firing positions and prepare to repel any attack. The real challenge, however, was quickly apparent: the hill was overrun with rats. They were everywhere. If you slept on the ground with your air mattress, you'd feel them scampering across your feet and legs all night, searching for food. On our second night, I decided to string my trusty jungle hammock between two posts over a trench. Still, I was awakened in the night by my poncho liner being pulled off me. Looking down, I saw a huge rat trying to use it as a ladder.

There were many similar incidents that night. As a result, we held a leader's meeting and declared war on the rats. Prizes were offered daily and overall, for the highest "body count." The only rule: you couldn't use any official weapons (no M-16s, pistols, grenade launchers, grenades, etc.). Creative traps were encouraged.

We hadn't counted on the ingenuity of the 101st troopers.

Along with classic box traps and snares, they rigged up booby traps with blasting caps and firing devices ("clackers") originally designed for Claymore mines and C-4. They baited the traps with peanut butter, turkey, jelly, cracker paste, and other food. Some soldiers even watched their ambush sites with night vision scopes and extended wires to reach a broader area.

Each morning, the results were displayed with piles of rats in front of every bunker. There were a lot of rat casualties, and even more amused and happy troopers, who enjoyed the distraction from our usual deadly business in the field.

I recall this story as a testament to the American soldier's ability to improvise and adapt. And, of course, after that first night, we ordered more blasting caps.

LIFE UNDERWATER

This is another example of GI ingenuity. Camping out during a typhoon was a unique experience. We were patrolling near Firebase Rifle during monsoon season. The area was thickly wooded, so we were searching for mortar and rocket firing positions or any sign of enemy activity. Rifle was far enough out that there could be NVA as well as VC threats. My platoon set up a patrol base along a trail running on a small ridgeline with steep drop-offs at both ends. Dense trees and underbrush bordered the trail, which sloped away on each side. We ran squad-size patrols for a day before the typhoon hit.

A typhoon—what they call a hurricane in the Southern Hemisphere—rolled in, and with no possibility of extraction, we had to ride out the storm, which seemed to last for days.

I assigned squad sectors for defense with our M-60 covering the trail into the deeper jungle. Poncho shelters were allowed for the soldiers not on watch. The standard setup—a tarp of poncho strung between two trees or poles—became our home. After hours of relentless rain, so heavy you couldn't see a hundred feet down the trail, the platoon sergeant and squad leaders all agreed: no enemy in their right mind would be out in this. They were sure to be holed up in a dry cave or bunker. With that, we let the troops focus on building shelters, keeping security on the trail, while drawing the perimeter in tight on the ridge for better drainage.

The shelters themselves were brilliant. Most soldiers had two ponchos, so a fire-team of five or six could snap and tie ten to fifteen ponchos together, making a "hotel" with a floor. It was quite a sight—and a huge morale boost. We could wring out our clothes and, with body heat, dry them enough that we didn't feel like we were in the Arctic rather than the jungle.

I remember an NCO lane grader speaking to my Infantry Officer Basic Course at Fort Benning as we prepared for a 3–4-day field exercise in constant rain. He said, "I've done this in Vietnam — for 45 minutes it's fun, after 2-3 hours it's really uncomfortable, and after half a day it feels unbearable. But I know you can do it, because I have." He was absolutely right.

So, the next time you get caught in a rain shower and get soaked running to your car or building, think about the infantryman who has to stay outside and keep fighting — when it's not just uncomfortable, but unbearable.

LARRY J. GARLAND

BRANCH OF MILITARY AND JOB HELD: Army, Military Police, Platoon Leader

DATES OF MILITARY SERVICE: 8 August 1969 – 18 August 1996

UNIT(S) SERVED WITH IN VIETNAM: B Company, 720th MP Bn, 18th MP Brigade

DATES SERVED IN VIETNAM: 14 August 1970 to 1 June 1971

HIGHEST RANK HELD: LTC USAR (O5)

PLACE AND YEAR OF BIRTH: Murray, Kentucky – 1947

ALL AMERICAN: MISS AMERICA 1970, PAMELA ELDRED

> "Put your hand on a hot stove for a minute and it seems like an hour. Sit with a pretty girl for an hour, and it seems like a minute."
>
> — Albert Einstein

That's a quote that has more than a little truth in it. I can relate my experience with four beautiful women touring Vietnam that required security escorts, and the hours spent with them seemed much longer. In mid-August of 1970, Miss America did a USO tour for 21 days, with one of the shows held in the Can Tho city area.

The beauties consisted of Miss America, Pamela Eldred from Detroit, Michigan; Miss California, Susan Anton; Miss Tennessee, Mary Cox; and Miss Alabama, Ann Fowler. Security for a USO show is generally about crowd control and protecting the USO members and celebrities. Any military person who could get the time off from unit duties was usually excited to attend any USO show. The audience was always predominantly male, and all

Western women were idolized. The troops loved and missed their round eyes.

I honestly don't remember what Miss America, or any of her companions, did for the entertainment part of their show—did it really matter? I was too busy observing the crowd and reviewing our assigned MP positions. I know what you are thinking. It was tough duty, but someone had to do it! Miss Eldred's talent was ballet, and she demonstrated that skill set with dance in her previous state competitions: Miss Detroit pageant in 1969 and the Miss Michigan title.

In researching a *Newsweek* interview with Miss Eldred after her tenure of service, she presented an interesting comment on the question asked of her during her competition. The following is a quote from the article:

In the '60s women didn't have the platform we have now, but during the pageant, I was able to create a platform sort of by accident. I was given a biography to fill out, so I wrote my siblings name and ages. In those days, the final questions weren't that difficult. My question was, *What advice would you give your sister, Melanie, entering the "now" generation?*

Pamela stepped up and said, 'Melanie was unfortunately brain damaged, so she would never enter the "now" generation, but that I would encourage any other young girl to enjoy life by being themselves.' It was just a question I answered. I didn't expect that much attention. But it shocked people and got a big reaction. I was then able to do work talking about people with learning disabilities during my time as Miss America.

That honest heartfelt answer was a demonstration of her moral character and won her the crown.

The show ended, and our mission continued to the hotel quarters with the USO group. We were tasked with the responsibility of security of the premises. We had guards posted at all entrances and changed guards on four-hour intervals.

The surprise part of the evening came in the form of a request from the tour manager. Their next day was designated a travel day and had free time embedded. The women had decided that they wanted to go shopping in the Can Tho market.

You can validate with my wife: shopping is a genetic characteristic in women that requires at least three outings a week, even if purchases are not made, but better if they are.

We took the pageant women to the open market in Can Tho the next morning. The minutes did seem much longer as we continuously scanned the crowd and area for threats as the women searched for Vietnamese wear, shoes, and other souvenir items. We were most relieved when the tour manager said it was time to load up, and the group departed the area in their buses.

The Miss America tour was a most pleasant group of people, but my memories were of anxiety and stress in ensuring that nothing would go wrong.

Anxiety and stress. That pretty much summarizes any soldier's memory of their duty in Vietnam. Every soldier serving in Vietnam has a different story to tell. Give many thanks and much respect for those that served in combat missions. They deserve so much more.

ALAN GRAVEL

BRANCH OF MILITARY AND JOB HELD: US Air Force Pilot

DATES OF MILITARY SERVICE: June,1969 through June, 1974

UNIT(S) SERVED WITH IN VIETNAM: 536th Tactical Airlift Squadron, 483rd Tactical Airlift Wing; 4102nd Aerial Refueling Squadron (Provisional)

DATES SERVED IN VIETNAM: C-7 Caribou – September 1970 thru September 1971; KC-135 – May 1972 thru December 1972

HIGHEST RANK HELD: Captain (O3)

PLACE AND YEAR OF BIRTH: Alexandria, Louisiana - 1945

RECOVERY OF C-7A CARIBOU KH 146

by Alan Gravel [536rd TAS, 70] and Jerry Pfeifer [483rd TAW, 71] from C-7A Caribou Newsletter Vol. 36-1, April 2025

Alan Remembers The C-7A Caribou Association Reunion 2023

I published my book *Haulin' Trash & Passin' Gas* about my adventures in Southeast Asia in the fall of 2020. For the cover, I chose a picture I took in Duc Lap the day I was on Maintenance Alert at Cam Ranh Bay (CRB). I chose that photo partly because it had good color, focus, and composition relative to many of the other pictures I took which were foggy, blurry, or unfocused due to the conditions when I snapped the shutter.

In November of 2023, the C-7A Caribou Association held its first reunion in several years in Kissimmee, FL. Previous year reunions had been canceled due to the COVID pandemic. I took a few of my books with me, thinking that there might be some interest among my fellow C-7A veterans. On the day of arrival, I received a request for a book, so I took one down to breakfast the next morning. I took a seat at a table with an attendee, with whom I was somewhat acquainted, and placed the book on the table to my left. A few minutes later, two other attendees that I did not know joined us at the table.

It wasn't long before the fellow to my right asked, "Where did you get my picture?"

I replied that I did not know what he was talking about. He indicated that the picture he was referring to was on the cover of my book. I told him, "I took that picture in Duc Lap in 1971." He pulled out his phone, flipped through his pictures for a few seconds, and then showed me a picture virtually identical to the one on the cover of my book.

At that point, it did not take a rocket scientist to figure out that I had been the pilot on Maintenance Alert, and he had been the maintenance officer who had repaired the damaged rudder on June 17, 1971, that was shown in the photo. Throughout the rest of the reunion, Jerry Pfeifer and I spent time together swapping stories. He filled me in on details about the Duc Lap repair that I did not know, and we also exchanged pictures of the event.

Jerry Pfeifer Recalls the events of 52 years earlier

June 16, 1971, began as a standard workday at Cam Ranh Bay (CRB) where I was the Wing Workload Control Officer. I had just returned to Maintenance Control from the Wing Commander's daily briefing when one of the dispatchers told me I had a telephone call from the Commander of the Special Forces (SF) camp at Duc Lap, near the Cambodian border.

I took the call and confirmed that aircraft KH 146 was ours. He said it had been hit by a Vietnamese Air Force (VNAF) H-34 helicopter attempting to ground taxi behind the airplane. The helicopter got too close and took a "big bite out of the rudder." He needed to get someone to Duc Lap as soon as possible to assess the damage and determine whether it could be repaired.

If so, the airplane needed to be repaired and out of there by dark the next day! He said the VC (Viet Cong) would drop in a couple of mortars for range the first night. The following night KH 146 would just be "a smoking hole on the parking ramp."

Such an attack would essentially shut down their fixed wing traffic, which was their main resupply source. The urgency of the situation got my attention. I got a couple of dispatchers on the line with me, and we started gathering information. Most of the damage was limited to the rudder and the left elevator.

Because Duc Lap was a forward base, their maintenance capability was limited to some crew chief hand tools. They only had one B4 maintenance stand. If they had to get to the tail of an aircraft, they did it with an extension ladder from the bed of a pickup truck. They did have several extension ladders, but nothing in the way of a crane or a high lift.

I told the SF camp commander I would get the wheels moving to get a team there with the appropriate specialists and would advise him when the maintenance team would arrive. Next, I briefed the Maintenance Control Officer and the Chief of Maintenance. They contacted Command Post to set up immediate transportation to take the evaluation team to Duc Lap. They also reserved an aircraft to transport the recovery team and equipment to Duc Lap the next morning.

Since I had most of the information, I was designated as the team chief and was told to get the team ready. I was experienced in such operations since I had worked on several C-130 recoveries during the previous five years. At the team meeting, I announced I had been named the team leader, gave everyone an update, and told them we had an aircraft available within the next hour to fly us to Duc Lap.

The Organizational Maintenance Squadron (OMS) said they would supply an experienced NCO to act as the overall project supervisor and two experienced crew chiefs. The Field Maintenance

Squadron (FMS) said they would send two experienced riggers to evaluate what needed to be done and the same riggers would also repair and rig the new controls.

The FMS supervisor said he had heard of some sort of bipod that could be used to lower/raise the rudder in emergency situations. It had been mentioned when they were inventorying the maintenance equipment for the projected transfer of the C-7A aircraft and equipment to the South Vietnamese Air Force. He said he would confirm the bipod's location and verify that it was complete and operational.

Finally, I gave everyone the name of one of my senior controllers and told them he was the point of contact for workload control. I asked them to keep him advised of any new developments. I also told everyone to identify people in the appropriate work centers to be on stand-by for any work that needed to be completed when the team returned that afternoon. The plan was to have all the repair equipment located and collected at the FMS hangar so it could either be loaded overnight or in the morning. I also asked all the team members to report to the FMS hangar the following morning to assist with loading.

At that point, I told everyone to round up their people and the equipment needed for the evaluation trip. We wanted to be airborne by 11 AM. I then gave them the tail number and location where we would meet our aircraft and told them not to be late.

We took off from CRB at about 1100 hours in a C-123 and arrived at Duc Lap about noon. From the air, it lived up to its "camp" name. It was a collection of sandbagged Conex shipping containers used as quarters, shops, and operational buildings, along with several temporary structures. Other than the short runway and parking pads, it was literally a "wide spot in the road."

So much for any embedded support! The telephone assessment was pretty accurate.

We spent the next two hours thoroughly inspecting the damaged areas of KH 146, particularly the damaged hinge points for the flight controls and the control cables. Fortunately, the aircraft had been unloaded before it was damaged. We used the pick-up truck/extension ladder method to get up to the hinge points for the rudder, the control attachment points, and to reach the top of the tail. We decided the pick-up truck and extension ladders would work in lieu of trying to transport more stands and equipment in, especially given our tight timeline. We then discussed and developed a plan of attack for the next day.

Before we left Duc Lap to return to CRB, I relayed an update to workload control via telephone and stressed the importance of finding the transportable lifting device. It was key to the whole replacement process and time was of the essence!

We arrived back at CRB about 1630 hours, at which time we got with the "home team" and told them what we had found. As we covered the information on the repairs that had to be done, people were taking notes and making lists.

The FMS supervisor reported they had found the bipod and determined it was complete — except for the two pads that were used to mount it to the horizontal part of the tail. He stated that they could obtain the bolt pattern, make two replacement pads, and weld them on to the legs of the bipod that evening. Once the replacement pads were attached, they would assemble the bipod to make sure they had all the required fittings and bolts.

When we finished the meeting, the treasure hunt began as everyone started to assemble the needed equipment.

By about 2000 hours that night, the new fittings had been made for the bipod, and it had been assembled. The bolt pattern had also been verified to match the actual attachment positions on the tail. We took plenty of cable to be used for guy wires and the lifting of components. Our transport for the next morning had been confirmed, so we accumulated all the equipment in the FMS hangar for loading and called it a night. We agreed to meet in the hangar at 0600 hours the next morning.

Everyone reported to the FMS hangar, and the loading went smoothly. We checked the load against our master list of components and equipment to make sure we didn't miss anything. In addition to my team, we also had the pilot (Alan Gravel) and his crew that had Maintenance Alert duty that day. They would stay with us and fly us back to CRB in KH 146 that afternoon. If we were not able to complete the repair in one day, then another C-7A would pick us all up, return us to CRB, and then fly us back the next morning — assuming KH 146 had survived the night!

Once we arrived at Duc Lap, the first two to three hours were spent getting our equipment in place and opening access panels. We also had to remove the elevator which had been damaged, which gave us access to get a ladder up to the hinge points of the rudder.

It was then time to assemble the bipod. This task was an exercise in brute force! First, we had to manhandle the two legs of the bipod up to each side of the tail, stand them on the horizontal surface, and then fasten them using the plates that had been welded on the night before.

Once the two legs were in place, we had a crew chief go up each side, meet at the top, and bolt the two halves together. A couple of guys steadied the unit on each side as the bolts went in and were tightened. We all breathed a sigh of relief when that task was successfully completed.

The rest of the job was a remove and replace exercise. Old parts down. New parts installed. No further flight-limiting damage was found and the repair flowed smoothly. The boom provided a simple, yet effective way to move the rudder into place.

As the operational checks were being made, we were all thankful that the Caribou had been designed as a tough, rugged, and basic aircraft that could "take a licking and keep on ticking!" Otherwise, we would have never met the short repair timeline.

The operational checks were all good, so we started breaking down our gear and loading it onto the aircraft. I talked with the camp commander and CRB maintenance and we all agreed a donation of the damaged parts to the camp dump was in order.

Alan Gravel's story as included in *Haulin' Trash & Passin' Gas*

Each day one crew would be designated for Maintenance Alert at Cam Ranh Bay. The alert crew would brief and pre-flight the airplane and then just sit around all day. If a flight crew had a problem with their airplane in the field, the Maintenance Alert crew would fly to the location with mechanics, parts, and tools.

After unloading, the operational crew would take the good airplane and continue their mission. The alert crew would wait around with the mechanics while the broken bird was fixed and then fly it back to Cam Ranh Bay.

I only pulled Maintenance Alert once during my tour. On that day, a South Vietnamese helicopter got a little too close to a Caribou that was on the ground at Duc Lap. The helicopter blade shredded the rudder but fortunately did not damage the vertical stabilizer, so the repair was a relatively simple parts swap that could be completed in a few hours. (*Alan did not realize in 2020 that the helicopter assault on the Caribou took place the day before.*)

Duc Lap was near the Cambodian border, southwest of Ban Me Thuot, and about 100 miles or so west of CRB.

We left Cam Ranh Bay with a new rudder, several mechanics, and a tripod device that they used to change the rudder in forward areas. We reached Duc Lap around noon and the other C-7A crew took our airplane to continue their operational mission.

While we waited for the aircraft to be repaired, I struck up a conversation with a SF advisor who had about 15 Montagnard troops with him. They had been waiting three days for transportation to Nha Trang. I offered to take them when we got the airplane repaired, as Nha Trang was just up the coast from Cam Ranh Bay, and we could drop them off on the way home. He checked with his people, and they agreed, so we did it.

LARRY GARLAND

ARVN MAJ NGUYEN — GONE BUT NOT FORGOTTEN

"Only the dead have seen the end of war."

— George Santayana

They were leaving: The 25th Infantry Division was programmed for redeployment during the 45-day period 1 November—15 December 1970. (Division OPORD 183-70 published on 15 October 1970)

The Division had a distinguished record in their AO: 1st Brigade in Tay Ninh, Bin Long, and Binh Duong Province. 2nd Brigade, east of Saigon in Phuoc Tuy, Long Khanh and Bien Hua Provinces. 3rd Brigade in Tay Ninh and Binh Duong Provinces with interdiction of enemy movements on the Saigon River Corridor. The Division eliminated 110 enemy during the month of October, accounting for 77 individual weapons, seven crew-served weapons, and .92 tons of rice captured or destroyed.

During the Division's deployment, they lost 4,654 American soldiers' lives. MAJ Nguyen, ARVN, was part of the Vietnamese III Corps that was designated to take over the Area of Operations, Fire Support Bases, and key installations.

I arrived at Cu Chi in early 1971 to claim three buildings to house MP Operations. I met with MAJ Nguyen, my counterpart, to introduce myself and overview our convoy missions. Major Nguyen could speak much more English than I could speak Vietnamese, and he suggested we go to dinner locally for additional follow-up on area operations.

I drove one of our trusty MP jeeps to a local outdoor café on QL4 (major highway) and met with the good Major somewhere around O'Dark thirty. We ordered what he said was venison and it came with potatoes and a little carrot. I just remember the meat did not taste like any venison I had ever tasted, but I ate it anyway.

In our discussions, I learned that there were many problems in the coordination of the Vietnamese operations after the 25th Division departure. Seems they were working to integrate their issues of concern and attempt better operational results. Most issues revolved around supply support from the U.S. military. I also learned that MAJ Nguyen had a wife and two girls. He had been in ARVN service four years and had primary duty of coordination of RF and PF personnel.

My unit, B Company of the 720th MP Bn, had mission responsibility for convoy escorts in regions III and IV, security of vital installations, and maintaining law and order under war time conditions. Running the platoon remotely from Company HQs was becoming a real challenge. We received remote support from other units for food and fuel. We always had our hand out.

The number of convoy escort requirements had been reduced in early 1971 and unit morale was suffering. Most of our unit MPs were just kids in combat boots looking forward to their separation date as a soldier. I can remember giving a situation report on reported failure of guard duty responsibilities and the consequences for lax security in our area. Several individuals were sleeping on duty or taking drugs. Yes — even MPs were having the same problems as other units. We were all just trying to "get home" with some sanity.

Early in our occupation of the platoon's new buildings, we discovered that the 25th guys had planted claymore mines around two embankments for security of their soldiers. Not having any knowledge of the mine placements was not good. I decided to drive over to MAJ Nguyen's unit to determine if any maps had been kept on the mine placements. Upon arrival at the HQs, I asked for MAJ Nguyen's location and was met with a rather curious look, perhaps angry expression.

After finding someone who could speak English, I was informed that Nguyen had been killed on his travel home two nights back. The details were very sketchy but apparently, he had been ambushed while entering his home neighborhood. It just wasn't my place to determine the specifics or possible motive. It was a shock, and I almost forgot about the task of finding a map for the claymores.

Upon parking the jeep back at our quarters, I heard loud music coming from the main barracks. If I had heard this song once, I had heard it a hundred times. It was the Animal's Band and the refrain "We got to get out of this place, if it's the last thing we ever do."

I was in complete agreement.

ROBERT "BOB" HOUGHTON

BRANCH OF MILITARY SERVICE: Navy Reserve

JOB HELD: Property Book Officer for Navy Advisory Group

DATES OF MILITARY SERVICE: Active Duty: August 1970 to September 1974; Reserves: 1974 through 1996; Total 27 Years

UNITS SERVED IN VIETNAM: Navy Advisory Group

DATES SERVED IN VIETNAM: 11 August 1971 to 9 August 1972

HIGHEST RANK HELD: Captain (0-6)

PLACE AND YEAR OF BIRTH: Chicago, Illinois — 1947

LETTER TO MY PARENTS

I don't think my situation was unique from others who served in Vietnam. As an only child, my parents didn't want me in Vietnam. They couldn't understand why I volunteered to go to the War.

What might be unique is that the way they raised me was a primary reason for me choosing to volunteer for duty in Vietnam. I am adopted by two late in life parents. My dad served in Europe in WW1 and volunteered to work at the draft board two nights and Saturdays during WW2. Certainly, after sending thousands of young men to serve, he should have understood.

My mother was active in the American Legion Women's auxiliary. My earliest memories were on Memorial Day when we would participate in laying wreaths on Veterans graves and my mom singing the National Anthem loudly and beautifully.

Below you will read the letter I wrote to my parents explaining my decision and beginning the transition from child to being the leader of our small family. Like many fellow veterans, I found the letter many years after returning home. With tears in my eyes, I read the letter. I was surprised in my words and my tone as I lectured them on their role in my decision.

My letter begins with a "White Lie" that my orders had been changed and that I was going to Vietnam. I had the orders for several months but hadn't told them.

Here is a slightly edited version of my "Letter to My Parents."

I know this doesn't make you happy, but I hope you will understand. If you think about how you brought me up, I hope it will make you proud, rather than sad. I feel that based on the beliefs you installed in me, it would be very hypocritical not to do what I am doing.

There are many things wrong in our country today and will only get worse when people like you and me don't accept the challenge being thrown at us. If you are upset, think about the millions of other parents worried about their children serving in Vietnam. If we want changes made, we can't wait for others to make them for us.

I do not have all the details yet, but I feel reasonably sure that I will be in a safe position in Saigon. I have talked to people on the staff here who served in Vietnam and received information from people over there now. Life in Saigon can be compared to any other foreign big city.

There are nine people from my Supply Corps class with orders to Vietnam, all, like me, volunteered. People who volunteered seemed to have many things in common with me. My roommate in Athens and two of my best friends are going, and we all have assignments in Saigon.

I have been very influenced by the Quast family, specifically Phil Quast. He had just gotten back from Vietnam and introduced me to many people he served with. I was very impressed by their pride and satisfaction of serving in Vietnam. I felt after meeting them and weighing the pros and cons, it was the best choice for me.

Of major importance to me is the hypocrisy I have already discussed. I would have felt bad with any other decision. I am very curious to learn what is going on over there. There are so many conflicting stories. I don't think the protestors are right, but I want to learn for myself.

There are financial benefits, higher pay, less taxes, and 10% savings added to my pay. I would have to have a job paying double what I made in sales for Olivetti, the job I had before going to OCS.

If I was assigned to a ship near you in San Diego, there is a good chance I would be deployed for seven months or more. When I get back, I can choose shore duty in America or a foreign country if it is reasonable and available. In the long run, I may be reducing the amount of separation from you.

There is the possibility of travel on R&R to Australia or to Japan and the Winter Olympics. I can also do a week in Hawaii where you could join me. I don't expect to be in a combat position, but in a safe position in Saigon.

I hope this letter will assure you that I have thought this out. I have been thinking about it ever since staying with Phil Quast on the weekends at OCS. I have had many close friends, family members, fraternity brothers, and our neighbors in Park Ridge who have served and returned safely home.

So far, I have been very happy with my life in the military. I feel I am in a much better position than my friends who aren't serving our country. I feel I owe nearly all the credit to you. You taught me a lot of things my friends don't know or appreciate… what you did for me growing up. I realize the sacrifices you both made for me, and I only hope that someday I can give myself to someone as completely as you have given yourselves to me.

I know your immediate reaction will be one of concern. I hope you realize that this was an important decision for me to make, and that I will need your support more than ever during the next year.

I look forward to being home before I leave, all the training will be in San Diego. I don't leave until August 19, nearly a month from now. If you have questions, I hope you will talk to Rev. Huls. He has served in Vietnam and would be a good person to talk to.

I know you love me a lot. When you sit down and think about it, I hope we all feel I have made a good decision

Love, Robert

I am grateful my parents kept the letter. We never had a meaningful discussion about the letter, but it seems it was important for them to keep.

My orders were modified for an earlier deployment on August 11, 1971. I remained in Saigon as Property Book Officer leading a unit of a dozen men and millions of dollars of inventory. I relieved Doug Edwards, an Atlanta Vietnam Veterans Business Association member. We have maintained a relationship with Doug and his wife Betty.

I lived for most of my tour with my roommate from Navy Supply Corps School. I didn't make it to Australia, but I did get to go to Japan and the Winter Olympics in Sapporo with another Supply Corps classmate. I also spent R&R in Hawaii with my parents.

Most, if not all of the nine people who went to Vietnam with me, didn't spend the full year, but I was there 364 days and came home on August 9, 1972. I was never in danger while in country. I feel I grew as a person and value the experience I had leading people and providing important supplies for Navy Advisors in the field.

BARRY PENCEK

CHRISTMAS IN THE JUNGLE

It was late December 1970 and the forecast in I Corps showed no snow for the holidays. Since a White Christmas was not in the cards, the rotor-heads of HML-167, based at Marble Mountain Air Facility just east of Da Nang, wanted to do something to bring a bit of holiday spirit to the grunts in the bush.

As fate would have it, Bob Hope was in Da Nang for his annual Christmas show and the Huey squadron wanted to welcome the comedian with a flyover and perhaps get a few seconds of fame on television. An aircraft was painted up like a candy cane for the occasion with "Ho Ho Hope" written on the side.

Once the USO show moved on to its next venue, it was decided the beautiful holiday paint job shouldn't be wasted and beer runs for the grunts, complete with a crew chief dressed as a white-bearded Santa Claus, were scheduled.

I was providing the gunship escort for Santa in a Cobra that day and it turned out to be one of the most fulfilling Christmases of my life. I'd check in with small OPs, radio relay sites, or artillery positions, and ask how many Marines they had. If they said ten, Santa would swoop in with a "ho, ho, ho" and give them ten ice-cold beers. The grunts couldn't believe what they were seeing, and I couldn't stop smiling.

People sometimes, tongue-in-cheek, throw about the cliché "It's the thought that counts." For me, Christmas 1970 was a great example that gift giving isn't about what someone gets, or how much it costs. The most important thing truly is the thought behind the gift.

PAUL WILLIAMS

BRANCH OF MILITARY AND JOB HELD: US Air Force, Acquisition Officer
DATES OF MILITARY SERVICE: May 31, 1956, to August 1, 1991
UNIT(S) SERVED WITH IN VIETNAM: Detachment Air America 11, Air Force Contract Maintenance Center, Udorn Royal Thai Air Base, Thailand
DATES SERVED IN VIETNAM: 7 July 1970 to 9 July 1971
HIGHEST RANK HELD: Colonel (O-6)
PLACE AND YEAR OF BIRTH: Minneapolis, MN — 1938

WORKING WITH AIR AMERICA

I was assigned to the Air Force Contract Maintenance Center at Wright-Patterson Air Force Base. This organization oversaw detachments located at contractor facilities worldwide. As part of the initial team who were members of this organization, we developed operating procedures, conducted staff visits, and ensured consistency in our operations.

After my first year, I found serving in a detachment to appear more rewarding than being in the headquarters. When visiting detachments, I was able to understand operations and the impact of directives. My goal was to serve in Southeast Asia, and my career advisor said my chances were good, but with my specialty, there were only a couple of areas for me to serve.

In June 1970, I was assigned to Udorn, Thailand, overseeing contracts for Air America. My responsibility was to administer the largest dollar value contract for maintenance in Southeast Asia, valued over $10,000,000 per year. There were many changing requirements which necessitated amendments to contractual arrangements. Negotiations were continual to make sure the customers were happy, and their needs met. Communications with various parties were good, considering the distance between us.

Air America inserted and extracted US personnel, supported the Royal Lao Army and Thai forces, transported refugees, and conducted photo reconnaissance missions. These were early operations as US military involvement in Southeast Asia grew. Civilian-marked aircraft controlled by the Air Force often rescued downed US pilots. Air America pilots uniquely operated non-FAA certified military aircraft in combat roles as private corporate employees.

Our effort and services were performed at Udorn, Thailand, Vientiane, Laos, and Phnom Penh, Cambodia, as well as various sites. Most of these sites had small landing strips, and we provided maintenance and assistance as required.

Seventeen aircraft, both fixed-wing and rotary, were dedicated to operations in Cambodia, Thailand, and Laos. In 1970, Air America delivered 46 million pounds (21,000 metric tons) of food in Laos, with helicopter flight time exceeding 4,000 hours a month.

Air America handled all levels of maintenance, from routine to complex tasks. I worked with talented colleagues… once, during an inspection, I found damaged metal on an aircraft, and our team expertly assessed and replaced the fuselage material, then I approved the repair. I saw many more similar repairs and was happy to have this expertise in an area such as this.

Air America had several aircraft in depot repair and the expertise in doing these repairs was amazing. The skilled workforce could handle anything from a simple repair to those depot tasks. When inspecting completed tasks, requesting rework was almost unheard of, and pilots would decide whether the repair fixed the problem.

I managed quality assurance for fixed wing aircraft, including inspections and arranging pilots for check flights after maintenance. Many pilots found it enjoyable to fly non-frontline aircraft such as T-28, O-1, and C-123, providing a break from their regular non-flying duties.

The contract with Air America was prepared and negotiated by the Sacramento Air Logistics Center in California. Coordination of activities was maintained with the nearby joint military assistance group. The main detachment of our organization operated out of Bangkok, allowing for collaboration with multiple sources. I received directions from many sources and was fortunate to keep everyone happy.

By mid-1970, the airline had two dozen twin-engine transport aircraft as well as Boeing 727 and Boeing 747 jets, plus two dozen fixed wing short-take off-and-landing aircraft, in addition to 30 helicopters dedicated to operations in Burma, Cambodia, Thailand, and Laos. There were more than 300 pilots, copilots, flight mechanics, and airfreight specialists based in Laos.

Air America flew civilians, diplomats, spies, refugees, commandos, sabotage teams, doctors, war casualties, drug enforcement officers, and even visiting VIPs all over Southeast Asia. Part of the CIA's support operations in Laos involved logistical support for Hmong militia fighting the North Vietnamese forces and their Pathet Lao allies.

Thousands of tons of food were delivered via Air America routes, including live chickens, pigs, water buffalo, and cattle. On top of the food drops (known as "rice drops") came the logistical demands for war itself, and Air America pilots flew thousands of flights transporting and air-dropping ammunition and weapons (referred to as "hard rice") to friendly forces.

My service ended in June 1971, and I returned to the United States. The year was eventful, and I served alongside many individuals. I am proud to have been associated with this company, its people, and my Air Force colleagues.

DONALD H. NAU

BRANCH OF SERVICE: U S Army – Armor/Infantry Officer
YEARS SERVED: 1967-1974
UNIT SERVED WITH IN VIETNAM: HHC, XO/CO, 3rd Combat Brigade, 25th Infantry Division
DATES IN VIETNAM: Aug 1969 to Aug 1970
HIGHEST RANK: 1LT (O-2)
PLACE AND YEAR OF BIRTH: East Cleveland, Ohio – 1944

"30 DAYS! – YA GOT 30 DAYS!!!"

Bellowed the Army Staff Sergeant. I was in a group of new in country replacement arrivals to the 25th Infantry Division's "Welcome to Cu Chi Orientation". We were seated on very old, rotted, splintery, and rough wood posing as former bleachers — riddled with years of weather, shrapnel grooves and slices. Paint? — never had it.

"THE FIRST 30 DAYS!!!," he continued in his loud monotone holler, "will be as dangerous as your LAST 30 DAYS." First, allow me to explain: On sign-in day, we then knew the exact day (365 days later) we would leave 'Nam to return to The World (USA). In between, there would be the WAR!

Somewhat calming down — he DID have our attention — the Sergeant condescendingly told us we knew nothing. "Knowing nothing," he went on, "would get you and others killed." I recall he never said wounded. "Listen to your NCOs," he barked, "AND anyone else who has been here more than 30 days."

"BUT, yet even worse," he continued, "Your LAST 30 DAYS will be just as bad. You will think you knew it all, seen it all, and would be counting down the days—aiming for that single-digit midget status."

He was right—about both.

Our Brigade Chemical Officer, MAJ Kinsey, had seven days to go for the "Freedom Bird" to take him home. Because his replacement had just arrived and was in his first 30 days - Kinsey went on the mission. After all, he had done many, many, many before. The flight included the hand-held spraying of a chemical agent on the growth around enemy tunnel openings. Interestingly, Kinsey would bring back the empty canisters as they could be used by "Charlie" in booby-traps.

Kinsey's Huey was shot down—all killed. As CO, I inventoried his personal effects. In the process, among some family photos, I found "empty" canisters of a defoliant chemical agent. Funny thing though, they weren't empty. Like a Coke or beer can, they still had a little left in them. Thus, the remaining liquid agent flowed down my hands and arms. I remember laughing—because I recall those "empty" beer cans in college. My Frat Bros and I would lay them down and, in a bit, there would be the last drops—yeah, the last.

The VA has recognized my direct exposure to Agent Orange with a diagnosis of squamous cell carcinoma—cancer. I have had surgery at the VA. Continual follow-up is ongoing.

Perhaps that exposure is not connected to MAJ Kinsey's last 30 days—a mission he did not have to go on. He did because he knew the job—well AND, his replacement was in his first 30 days.

RICHARD DITTMAR

BRANCH OF MILITARY AND JOB HELD: Army, 11E2T – Armor Crewman
DATES OF MILITARY SERVICED: August 1969 – February 1972
UNIT(S) SERVED WITH IN VIETNAM: 11th Armored Cavalry Regiment
DATES SERVED IN VIETNAM: January 1970 – December 1970
HIGHEST RANK HELD: Sergeant E5
PLACE AND YEAR OF BIRTH: Rolla, Missouri – 1951

IN MEMORY OF PETER STADDON

Staddon came to K Troop, 3/11 Cavalry Regiment in Vietnam late in 1970. In the few weeks that we knew him, the impression he gave was of a smiling, good-natured kid. He seemed to enjoy the light hazing we gave him and was never seen without a big smile on his face. At that point of my tour in Vietnam, I was a gunner on an M551 Sheridan tank.

We would patrol (bust jungle) by day and return to our troop base camp at night. On October 30, 1970, due to understaffing, our platoon was shorthanded and under gunned. Instead of our usual complement of three Sheridan tanks and four ACAV's (Armored Combat Assault Vehicles), we had two Sheridan's and three ACAV's, each with no more than three crew members.

As night fell, we were unable to return to our base camp and instead formed a circle facing outward between a raised highway and a nearby tree line. We settled in for the night. Later in the night, we came under heavy small arms, machine gun, and RPG fire from both the highway and the tree line.

During a fierce firefight, we received close air support with two Cobra helicopter gunships arriving with clear lines of fire on both sides. A platoon of M Company large tanks (M60's) arrived on top of the highway and set up a defensive position. As daylight broke, a platoon of ARVN infantry and other support arrived.

Staddon, sleeping inside the platoon commander's ACAV, was killed instantly when the first RPG struck and exploded inside.

A few days later, I received some photographs from a roll of film that I had sent for processing a week or two earlier. Inside were some photos of Staddon, each with his big smile and bright demeanor. I resolved that I would contact his family when I returned home to share these photos with them. However, the Army sent me to Germany after a short leave. When I left the Army in early 1972, I forgot my resolution and returned to school, then started my career and family. In 2010, forty years later, I rediscovered those photographs and was reminded of my original plan.

By then, I had the advantage of the internet and began searching for his family. Since Peter was a Mormon, I began my search in Utah, but could find no one with the surname Staddon. A fellow trooper I had stayed in touch with said he thought Staddon was from Arizona. I finally reached his grandmother there. I told her who I was and why I was calling. She understood and gave me the phone numbers of Peter's two brothers. I was able to reach one brother and explained my purpose. He was gracious and related what had happened after the loss of Peter.

His mother, devastated by the loss of her son, fell into a deep depression and did not recover. On the following Veterans Day, she took her own life.

VIRGINIA L. "GINNY" DORNHEGGEN

BRANCH OF SERVICE: Army Nurse Corps — Intensive Care Unit and Recovery (PAR) Nurse
DATES OF SERVICE: 1970-1973
UNIT(S) SERVED WITH IN VIETNAM: 67th Evacuation Hospital, Qui Nhon
DATES SERVED IN VIETNAM: Nov 1970 — Nov 1971
HIGHEST RANK HELD: 1LT (O-2)
PLACE AND YEAR OF BIRTH: Gettysburg, PA — 1948

GINNY'S STORY

I always knew I wanted to be a nurse!

As a young woman, before I was married, I was Ginny Deardorff. I was in the Girl Scouts, was a Candy Striper at the local hospital, and after taking an American Red Cross course, I babysat. I loved the adventures of scouting (camping, backpacking, travel), but most of all, I found a great deal of satisfaction in volunteering to help others. And, that's probably a prerequisite for someone considering nursing as a future profession.

Back in the 1960s, a three-year nursing school program was an option for year-round learning and that's what I chose. Growing up in Gettysburg, Pennsylvania, I attended the not-too-faraway Harrisburg Hospital School of Nursing. There, other students and I stayed in a dormitory, worked long shifts at the hospital, and took classes between shifts.

Wanting to expand my horizons after graduation, I considered possible options. An upperclassman told me about the Army Nurse Corps (ANC). It promised travel, additional education, pay the senior year of nursing, and only required a two-year service commitment.

Intrigued by the opportunity, I remember going home and telling my parents that I wanted to join the Army Nurse Corps.

I was very excited and told my parents about all the things ANC could offer me. I wanted to do something special and meaningful with my life, while learning at the same time. But, most of all, I wanted to serve my country and continue the sense of pride and loyalty I was taught in the Girl Scouts. So many of us wanted to serve our country and do something meaningful at the same time.

With my parents' blessing, I applied to and was accepted into the Army Student Nurse Program my senior year. My Dad was immensely proud as he had served in WWII.

After graduation from the Army Nurse Program, I continued a postgrad course in ICU/CCU (Intensive Care Unit and Cardiac Care Unit). I believed that advanced training would better prepare me for future challenges I might face.

My first assignment was at Walter Reed Army Hospital in the Recovery Unit. There, I experienced firsthand the horrors of war wounds and learned the meaning of working long, hard hours. I accepted my assignment as preparation for the possibility of becoming a part of the war in Vietnam.

When I was considering Army nursing as a career choice, the recruiter told me that there was a 95 percent chance I would be sent to Vietnam. I'm not sure why, but I didn't really think that would happen.

* * *

After nine months at Walter Reed, I received orders directing me to service in South Vietnam. When the orders arrived, I read them, then cried. But, while I wasn't happy about going to a place where I might be killed, somewhere deep inside I knew and had accepted that possibility. I understood that the skills I now had would be valuable there.

Just like all of those who had gone ahead of me, I experienced an abrupt and dramatic change in my life when I arrived in Vietnam in November 1970. I had just turned twenty-two years old. Arriving as 2nd Lieutenant Ginny Deardorff, I was assigned to the 67th Evacuation Hospital in Qui Nhon, about 235 miles north of Saigon. Because the area was in a designated "Combat Zone," I would no longer be wearing a nicely pressed nurse's uniform.

Instead, I was told that jungle fatigues and combat boots would be required dress. After collecting my new uniform, I was immediately assigned to the PAR (Post Anesthesia Room — the same as a Recovery Room in the States) and ICU, two critical care areas where my training and experience would be put to work.

There was no 'Front Line' in Vietnam, I learned that very quickly! At night, bright-red tracers from M-16s, loud explosions, 'Red Alerts,' and the eerie light of illumination flares, were constant reminders that I was very much a part of the war in Vietnam.

As a 'Combat' nurse, my duties, skills, and expectations were to maintain the highest standards of care during any type of situation. Critical thinking skills were of utmost importance. I was trained to visually assess and listen so that I could quickly react to the needs of our soldiers. In Vietnam, there was no computer, iPhone, Wi-Fi, or Google to help with references—you had to use what was in your head.

Our soldiers were the reason that all the others and I were there. Our goal was to ensure that every injured soldier received the best of care and respect from every nurse tending to him.

Since a family member couldn't be there with them, we would hold the hand of a dying soldier to the last moment of death. We also wrote letters for those who had arms amputated or injury to their eyes. We shared stories, tears, and smiles. Our soldiers were young, strong, and truly terrific to care for. Most of their injuries were life-changing. And, caring for them changed my life.

In 1971, I was promoted to 1st Lieutenant. Many things happened as I continued to care for our soldiers and I could easily tell many stories, but this is my most-favorite story.

One day, Miss America, Phyllis George (and her entourage) visited the 67th Evacuation Hospital. She was quite beautiful with perfect makeup, beautiful hair, and she took the time to speak with each soldier in the ICU.

I was caring for a soldier who had both of his arms amputated when Phyllis walked over to speak to him. I helped by holding him up so he could see the American beauty better. Phyllis spoke with him several minutes and he had the biggest smile! As I laid him back down in bed, I said, "Wouldn't it be great if all of us nurses could look like that?"

Then the soldier looked into my eyes and said, "She doesn't hold a candle to how you nurses look and what you mean to us!" I immediately teared up, leaned down, and gave him a kiss on his forehead, and told him that I would never forget him. And, as you can see—I haven't.

Doctors and nurses were typically scheduled to work 12-hour days/six days a week. When Mass Cals occurred, it was more like 16-18-hour days. "Mass Cals" was an abbreviation we used for Mass Casualties; when fifty or more wounded patients arrived in a very short time. During the period the TET Offensive lasted in early 1968 (before I arrived), the 67th Evac sometimes had Mass Cals of more than 150 patients.

The camaraderie and high degree of professionalism displayed by the nurses and medical personnel I worked with were key to the 67th Evac's esprit de corps. Joining the Army Nurse Corps was the best career choice I ever made. The teamwork and willingness to help each other, no matter how busy, was my very favorite part of military nursing.

Now, well after fifty years since I served in that faraway place, I am Ginny Deardorff Dornheggen. I am a Vietnam Veteran, a survivor and souvenir of that War. I have chosen to remember the good things that happened, even though I do have anxieties remaining. I am proud of the privilege to have served my country with the young men and women I got to know, and — they will always be a piece of my heart.

DONALD H. NAU

ENTERING BAO TRAI

The dangerous road distance from Cu Chi to Bao Trai (pronounced Bow Try) was about 18 miles. The Army Jeep trip to the village took about 45 long, and highly alerted, minutes. Averaging 24 miles per hour, of course, meant sometimes faster or slower. Slower speeds were avoiding the obvious craters caused by road mines. Back in 1969, road mines were "planted" by hand and activated by a vehicle wheel rolling on it. As there were no North Vietnamese Army (NVA) anymore since the 1968 Tet offensive, the area was now ruled by both the Viet Cong (VC) and the locals who had to obey orders by the VC. During the day, the local families were rice-paddy workers — all waves and smiles to the GIs. At night, they would have to obey the VC.

For that drive, I wore a steel helmet and a vest (flak jacket) and had my regular .45 pistol and M16 rifle. In addition, I would have an Army 12-gauge shotgun across my legs with barrel pointed far out and showing. This would keep the locals, as well as children (who could carry a grenade) from approaching me. No one did—I never fired.

On most occasions, the entry into Bao Trai had a greeting with the line-up of bodies lying next to each other with feet at the dirt road. The bodies were physically torn, ripped with parts missing and chunks of flesh blown away. Splintered bones were showing. A lot were nearly naked—others had shredded black "pajamas." Many had the dead facial expressions with mouths and eyes open. Some didn't have a full face. There was a lot of blood drying on the dirt and their bodies.

Locals would walk by and not notice—they didn't look away—just not notice—this was not unusual—probably someone they knew. Admittedly, I, too, stopped noticing. The KIAs were a result of the Army of The Republic of Vietnam (ARVN). These were friendly and trained by the USA's 25th Inf Div. Their units surrounded Bao Trai and nearby Hau Nghia which was the Provincial Capitol. NOTE: Killing of South Vietnamese Leaders and Officials was a grand prize for the VC.

SO—Why go to Bao Trai at all? The 3rd Combat Brigade had responsibility to a large part of the 25th Infantry Division Area of Operation. That responsibility included the communications network maintained by my Brigade. As XO/CO, the enlisted personnel fell under me. Therefore, I would make, at least, monthly visits to "the men" for wellness, as pay officer and a meaningful pat on the back. The "commo" area at Bao Trai was in a truck near the US Tactical Operations Center. It was extremely important to keep this commo unit operating.

I cannot recall each day in Vietnam nor each mission, however, as I write this, I can vividly see, and cannot forget, the days I entered Bao Trai.

DAN HYDRICK

BRANCH OF MILITARY AND JOB HELD: US Army / 11B Infantry / 11C
Combat Radio /12B Engineer – 89D EOD - LRRP
DATES OF MILITARY SERVICE: 1969 to 1972
UNIT(S) SERVED WITH IN VIETNAM: Americal Division, 23rd Infantry,
196th – 26th Engineers, 75th G
DATES SERVED IN VIETNAM: Dec 1969 to Nov 1970
HIGHEST RANK HELD: Sergeant E-5
PLACE AND YEAR OF BIRTH: Wynne, Arkansas – 1951

> "Hello DARKNESS my old friend"
>
> "The Sound of Silence"

So, my wife asked me why I close the bathroom and the closet doors at night in our bedroom, as they both face the bed across the room. She said it's like a ritual that I do before going to bed. I check the perimeter of my home, checking the locks, setting the home alarm, and close the closet and bathroom doors that are in our bedroom.

I had to think a minute as she is a LPC (Licensed Professional Counselor) who has helped me throughout the years. I felt I should respond with a thoughtful answer. The answer is now my new project; clearing a 50+ years Vietnam issue that I forgot (subconsciously) that I apparently have carried for far too long.

Vietnam 1970 June, I think. We always said that we (Army) control the day, but Charlie (Viet Cong) controls the night. We were not wrong. If things are going to go bad, it will usually happen at night.

We were located outside of My Lai 2 in the Quang Nai province south of Chi Lai, under the watchful eye of LZ Dotty's artillery. It was a small NDP (night defense perimeter) that the engineers expanded, as we were told we would be there for a while. Part of the new Pacification Program was our mission. Winning the hearts and minds, as someone once said.

So, we dug new, half buried underground, sleeping huts made of sandbags and scrap metal. The opening to our bunker had a sloped entrance, as it was half underground. It slept eight of us with homemade cots to keep us off the dirt floor. The cots also made it harder for the rats to crawl on you.

It was a long day of digging, filling sandbags, and setting up new razor wire just outside the berm. We were cut up and tired as night fell. We played some cards, smoked a few Camels, and basically tried to remember if we tied off everything, meaning the razor wire, but the new sleeping bunker was pretty cool, but very dark and damp.

We hadn't slept for days, so I thought tonight would be a long hard sleep. But it was the darkest night I think I have ever witnessed, with no moon. Perfect time for Chuck (VC) to move around, so we were told we would be posting more guards and upping the rotation, meaning more guard shifts. I knew my sleep would be short.

I don't remember the time, just that I was hot and sweetly under my mosquito netting in our new home, half underground. I positioned my cot directly in front of the door opening, as there were no back door or windows. I felt less claustrophobic being by the opening, but it was really DARK!

For some reason, I laid my weapon just outside my mosquito netting as the bugs were eating us alive. I don't remember when I fell asleep, but it didn't seem long when something woke me up.

When I opened my eyes, there, standing in the doorway was a dark shadow. A small man carrying a strapped bag in his hand. Was this a Sapper? He seemed to be peering in and listening, trying to see into the bunker. He must have stood there for what seemed like an eternity, motionless.

I remained very still. I began rolling back my mosquito netting to get to my weapon. Every move I made seemed like it made him move, as if he noticed the movement. It was then I reached my weapon and slid it to me.

I quietly clicked my weapon off safety and moved it into his direction. I moved very slowly, hesitating to fire as I was afraid I would fire up a friendly, one of our own. I thought about calling out, but it seemed my choices were limited. If I called out and it was a Sapper, he would just throw what I believed was a satchel charge into the room that would surely take out all eight of us.

My heart was racing, and I needed to make a decision. My father, a boxer in his day, always said, he who hesitates, will mediate in the prone position. Now my shadowy friend was changing his position. It seemed like he was looking back out to the outside. He even moved just a little towards the outside. I still couldn't make out for sure who he was, someone coming to get me up for my guard duty shift, or truly the enemy.

It was at that moment he turned as if he heard something outside and ran off. Moments later, I heard the blast. I jumped up, tearing through my mosquito netting, and raced for the door.

The word went out, "SAPPERS in the wire!" All of us ran out as all hell broke loose. "They're inside the wire!" someone screamed out. It was apparent by the blast that I heard that someone had blown up a small trailer mounted tank right outside our bunker.

Now for those of us in my bunker, this gets real silly, lucky, or fortunate. You see that tank was marked FUEL, but it was really a WATER tank marked to confuse the enemy. As if the VC could read… however in this case, it just might have worked.

The firefight didn't last long, as we soon realized there was just one sapper, mine! Apparently, the sapper was shot while throwing his satchel charge not 20 feet from where I encountered him. Had it have been fuel, well, hell could not have been hotter.

I second guessed my choice not to fire almost immediately, re-living the ordeal again in my head. I was pissed that I did not light him up, (shot him). It did not make me feel any better that no one was hurt, they could have been. I began to wonder if I should tell the guys about my encounter or just let it go. After all, these guys would have had a field day telling me their opinions.

As I learned later that night, the irony of it all, was that we heard that the shooter, who took down the sapper, was none other than my platoon sergeant. He was bragging how he just so hap-pened to be coming to my bunker to get me up for my guard duty shift when he saw the sapper moving in the open.

I thought I was over my cautions about open dark doors, but now I have come to find out that I was not.

To this day, I cannot go to sleep with the closet or bathroom doors open. I cannot look at a dark opening at night without get-ting a creepy feeling.

Nam never leaves us, it just reminds us that Charlie will be with us forever, in the DARK! Goodnight.

DONALD H. NAU

THE BALL CAP

The recollection of my first ball cap goes back to about 13 years of age when I played on a school baseball team. I did not have one be-fore. My dad did not wear any ball cap, so I did not get his old ones.

However, as a right fielder on the school baseball team, my cap provided the sunshade for some great fly ball catches. Of course, it meant I was a team member, I was in, I was good enough. I must say the cap did not help when I batted. In fact, it was a nuisance when I swung with all my might, missed, and then had to pick it up from the dirt. Dirty ball caps are usually a badge of honor if you dove for home plate to score the winning run! I never did that.

A classmate in high school said, "Hats make you bald." So that was that! I Remember exactly which classmate that was. I know now he was no expert, science or medical student—and he and his family had the male balding tendency. Good enough. No hats.

My 2nd ball cap was not for sports. It came in the Army. Not in Basic or Advanced Training or even Officer Candidate School but, at Fort Hood in the 2nd Armored Division where I was a 2nd LT Tank Platoon Commander. When on Post, we would wear the olive drab ball cap—out in the field it was, of course, the Tanker helmet which had communication capability. In Vietnam, however, we wore a "Boonie" hat or a combat steel helmet (pot). Still have a Boonie today which is now relegated to mowing the grass while remembering a Vietnam favorite song "We Gotta Get Outta This Place" (an oldie by The Animals).

It wasn't until my wife bought me, for a Birthday gift—a Vietnam veteran ball cap or, rather, a VIETNAM VETERAN ball cap. She called it the magnet hat—she was right.

When the gift was opened, it was in a plastic bag, freshly made and clean and shiny. She then washed it many, many times. It took a while to wear it when shopping in town—she understood. When I did, sure enough, it would attract a conversation with someone who was either a Vet or a family member of a Vet who was living or deceased—some from 'Nam. Many conversations were with men who were my age and "wished" they had gone. Some were sent elsewhere for their "hitch", while others did not go into the service. I listened to all. Rarely, did anyone ask me what I did THERE—Whew......

Ironically, that magnet hat had an opposite reaction to what I could have ever imagined. It provided an opportunity for others to express their situation during those Vietnam times. Quite possibly, some of the stories had never been told to family members. I have encountered that phenomenon before. Somehow families are the last to know. The War was never discussed in my family. I was thankful for that. My father knew as he gave me the Blue Star flag he posted on the outside door. I came to believe my sister never knew I was there.

Now, it is not every time I wear the cap into the store. Rather, at times it is left on the seat of the truck where it, dutifully, awaits my return — similar to what dogs, horses, and Jeeps do.

Sometimes, I just don't want to think about The War.

Editor's note: I have one of those caps, which I switch back and forth with a 4ID cap every time I go out (must keep this bald head of mine protected from sun or cold). Most people do not recognize the 4ID patch, but they seem to always comment on the Vietnam Veteran cap.

VIETNAMIZATION ADVISORY PHASE: 1973-1975

JANUARY 27, 1973: Draft comes to an end.

FEBRUARY 12, 1973: POWs return to United States. First C-141 flight returns first American POWs to US as part of Operation Homecoming; an additional 53 flights returned over 2,000 servicemen by April 4, 1975.

MARCH 29, 1973: Last combat troops leave Vietnam.

AUGUST 8, 1974: President Nixon resigns as President of the US in the wake of the Watergate scandal. Gerald Ford sworn in as 39th President.

APRIL 30, 1975: North Vietnamese troops enter Saigon, ending the Vietnam war.

POST SOURCE: www.vvmf.org/VietnamWar/Timeline

Following is the longest story in the book. I didn't hesitate when I decided to break my own story length rules and include this whole story together. Read on and learn about the evacuation of Saigon on 30 April 1975 from a Navy officer who was on the ship that evacuated an untold number of refugees. You will enjoy it, as I did.

ROBERT SCHACK

BRANCH OF MILITARY AND JOB HELD: US Navy. Engineering Officer
A, M Division, Communications Officer
DATES OF SERVICE: 2/15/74 – 2/15/78
UNIT(S) SERVED WITH IN VIETNAM: USS Hancock CVA 19
DATES SERVED IN VIETNAM: 2/15/75 – 5/1/75
HIGHEST RANK HELD: LT (O3)
PLACE AND YEAR OF BIRTH: Glen Cove, NY – 1951

EVACUATION OF SAIGON

By April 19, 1975, we were back off the coast of South Vietnam. After we arrived, there was a whole lot of sailing around, and nothing going on. Sailing in circles. Being in engineering, you have no idea if you're off the coast of California or off the coast of South Vietnam.

One day, I told a second-class machinist mate who worked for me, Lee Boltz, I said, "Boltz, we've got to see what's going on." We headed up to the signal bridge, which is in the island above the bridge at the highest point you can get on the Hancock. It was a grey overcast day, and as far as you could see, there were ships. I just turned around in a 360-degree circle and saw up to the horizon, which is about ten miles from that height. I've never seen anything like this, nothing but US Navy ships of all types.

My impression was, "Oh, my God." You hear about how America is powerful, we're halfway around the world and there's almost 40 ships, I think, I can see from horizon to horizon. With that much power, you could only imagine what we could do militarily....

No instructions, no official briefing of what exactly was going on. Occasionally a helicopter would fly off to scout out what was happening in the vicinity of Saigon, one rumor, or fact, I can't confirm, was before all the action that was about to commence, the pilots on a Saigon flyover saw people still out playing golf! I'm a junior officer. At my level and below, we have no information other than rumors. We are going about our business and doing just our job... day after day. No news.

The next thing I recall is being on the hangar deck mid-morning when I realized something unusual was about to happen. I observed a large group of Marines, approximately 200 to 400 in the forward part of the Hangar Deck, dressed in civilian attire. Got their weapons, got their helmets, they're in civilian clothes. They flew the Marine troops on board into the country for security, to include Tan Son Nhut airport, which was in danger of being closed as the last flight option out of the country.

Those young guys were taking a lot of risk, if captured, and they're in civilian clothes, they're not technically being deployed as military combatants. I thought something must be going on. Saigon may be in trouble, and the evacuation is about to finally happen.

Next, among the Marine pilots in the bunkroom, young guys as I said before, they voiced their biggest concern that the North Vietnamese would resist the evacuation, and intelligence was aware North Vietnam now had portable shoulder fired heat seeking missiles to knock down planes and helicopters. New weapons that were not used in the Vietnam War, but in the 1973 Yom Kippur War, the Egyptians used them against the Israeli Airforce to great effect.

Flares were mounted on the Jolly Green Giants, since the missiles targeted a heat signature. The plan for the moment was for the helicopter crews to fire off these flares over the city as needed to avoid being shot down....

We then became aware that Tan Son Nhut airport was now

closed, via informal information circulating on board. The assembled fleet provided the only airlift capability. Of the four carriers deployed, USS Midway and Hancock had Marine helicopters deployed. Enterprise and Coral Sea were deployed with their traditional airwing for support. On Hancock, it was understood at this point that our role in *Frequent Wind* was planned to be evacuating US and critical dependent South Vietnam personnel and transporting them to other support ships in the fleet, as well as to us. We assumed it would be more like an expanded Phnom Penh.

There was a sense of calm and a little frustration. Since nothing had been happening for several weeks, there was no special preparation for what was assumed by the crew to be just another AirOp we needed to support, only with helicopters this time in place of the traditional airwing.

First go and airlift people from the Embassy and then bring them to other ships, then we would evacuate the Marines back to Hancock and we'd have a similar complement of people on board, maybe a little larger when the operation concluded.

Suddenly, out of silence on a grey, hazy, humid, overcast day. No wind. Still water. Time to go! It started. What was amazing to me over the next 4-5 days, my strong feeling was that in the Navy everything is planned, and you practiced for just about every contingency. But in this situation, there was no plan at all for the challenges we were to be faced with. I'm just proud to be an American, because there was no plan, no one had any concept of what we aboard would be contending with, yet everything came off without a hitch, despite having zero prior instructions. The crew responded and innovated on the fly to meet every challenge in a great display of unselfish teamwork. Officers and enlisted, it just came off right operationally.

Initially, our Jolly Green Giants took off in a planned sequence and proceeded toward Saigon. Both the hangar deck and flight deck were largely empty, as the Marines had already deployed several days earlier. The ship's closed-circuit television displayed live footage of activities on the Hancock Flight Deck, both fore and aft. Our Marine Pilots from the bunkroom had departed, and our self-installed television provided continuous coverage of all flight deck operations. All personnel, including the bunkroom Ensigns, were engaged in their respective duties; some were stationed on the flight deck, while others were either on watch or at their designated duty posts.

Next, unexpectedly, South Vietnamese military helicopters began to appear, Huey's primarily, loaded with families in search of a place to land in the middle of the ocean. Hancock's large open flight deck was an obvious welcome sight for the desperate people fleeing their homeland. A few landed on the flight deck unauthorized, it started like a trickle. Just a couple of Hueys. What is going on? They're just landing in a haphazard manner, the concern is, well, how are our other ships in the fleet doing, are they going to land on the other ones also? What are they going to do next, ditch in the ocean? How are our own Jolly Green Giants going to land on the trip back from the embassy?

Unexpected chaos from the scale of people fleeing to come out to the fleet with no expectation of surviving and short on fuel. What quickly became apparent was that panic has set in country and anyone who desired to flee is going to find a boat, or if you are in the military and have access to a Huey, you are going to take it, fuel it, load family and friends, and fly out to the waiting US fleet and try to land on something.

Next, two bi-rotor helicopters, Chinooks, also landed on the aft portion of the flight deck, and the order came down to clear the flight deck and get the passengers down to the hanger deck. So, they took the two Chinooks, and they moved to push them off the fan tail into the ocean! Humorously a big cheer goes up as the flight deck crew springs into action, enthusiastically thinking this is a onetime event. On an aircraft carrier, there is a four-foot mesh installed around the perimeter of the entire flight deck, to prevent anyone who accidentally falls off from falling all the way into the ocean. The flight deck crew hauled the Chinook to the fantail to push it over vertically, and the bottom landing skids got hung up in the safety mesh. It ended up hanging off the back of the ship at a steep angle, it was stuck at sixty-degree angle and didn't quite fall off.

They had to push them to a steeper angle, which did not quite do the job until gravity took over and the entire Chinook fell and tore the entire safety net structure on the stern part of the flight deck off. The second Chinook followed quickly off the fantail with no safety structure to impede its fall. Hancock, meanwhile, is underway and station keeping relative to other ships in the fleet.

Hueys loaded with families started to arrive with greater frequency. One of the bunkroom Ensigns on the flight deck had to jump into the safety mesh to avoid a helicopter landing on top of him... all watched live by on our little TV. The Vietnamese pilots were desperate, the people on watch on the bridge at the time told the story of one pilot who for whatever reason could not land, lowered his helicopter to about 15 feet off the water, pushed it to roll right as he jumped left into the ocean... I think he was picked up by Hancock's rescue helicopter.

Amazingly, in relative short order, a process was implemented to impose order on the chaos and allow the Vietnamese Hueys to land in a more guided fashion and provide space for our returning and departing Jolly Green Giants with their passengers throughout the day. It was almost like a factory production line. The helicopter would land, we had security people also deployed on the flight deck, security would get everybody out of the helicopter, they'd take the weapons that they had with them and throw them over the side.

Next, something which I thought was kind of dehumanizing, but they had some stuff they threw a handful on each one of the passengers, in case they had something wrong with them, disinfectant; Navy ships are very clean. These poor people, they're all really thin — men, women, children, babies, grandparents, military, they get off the Hueys, and their weapons were thrown over the side.

Some had big clear, plastic bags of white powder, the rumors were "that's heroin." I don't know what that was, but they threw those over the side, too. Then they would lead the people down to the hangar deck and would remove their baggage and give them a receipt. The baggage was piled up port side forward to be inspected and claimed later. As far as the helicopters, they stopped tossing them over the side indiscriminately.

Hancock had two main elevators on either side of the ship, they used the starboard side aft elevator behind the Island to lower the helicopter down to the hanger deck, inspect it, strip it, bring it across the hanger deck to the port side elevator, bring it up to the flight deck, and throw it over the side from the elevator, keeping the flight deck relatively clear.

The "production line" continued throughout the day of getting the people and helicopters that arrived unannounced processed; equally along with the people from the embassy disembarking at the same time. Everyone in the hangar deck was treated equally, no matter how they arrived onboard. The refugees in the front of the hanger deck, helicopter processing in the rear.

People kept coming, and coming, more than we ever in our wildest dreams anticipated. Ultimately 24 helicopters went over the side that day, to allow for more space and to accommodate more people. Vietnamese Hueys. They were all shot up…. We kept what we had room for that were in reasonable condition.

The people coming aboard were a cross section of life, these weren't just "select" people that we treated with a special priority, they were people from villages too. One thing that really impacted me was what I or anyone else would do if we were in their shoes and somebody said, "Hey, you've got two hours to get out of the country." What would you take with you? These people escaped with only what they could carry. The guys flying out to the fleet, how did they know they ever were going to find a place to land? Pure desperation.

When they arrived on board, there wasn't joy. More shock, relief, sadness, "Where am I?" bewilderment, "What's going on? Passive.

Meanwhile, we're still operating this ship, doing the mission, The Jolly Green Giants cycle back and forth between the ship and the embassy in Saigon all day, not distributing passengers to other ships per the original plan since now any ship with a helipad has Vietnamese helicopters trying to land on them.

Our Marine pilots were flying back and forth continuously and getting tired, but they kept on. One of the pilots who slept in the rack next to me in the bunkroom told me when I asked him the next day when the airlift was over what happened? How long were they on the ground? He described, "We landed at the embassy, (the helicopters were rated for maybe 45 people maximum), we landed, they opened the rear door, we have no idea how many people per trip, they weren't checking names off, they got in," they had a weight limit, and the Vietnamese were thinner, smaller statured people. They packed them in.

He said they landed, packed them in, and left. Rapid stop and go, just landing and taking off continuously. On the ground a very

274

short period, maybe transporting over 60 people per trip on average.

While the helicopters were cycling continuously over less than a 24-hour period, we improvised, and it was an amazing logistical thing. We had our helicopters landing, their helicopters landing. How we created space, how we cleared the flight deck, and we eventually started to stage the people as the day went on. The hanger deck was transformed into an ad hoc airport waiting room.

They created "squares" of refugees in the forward portion as they were processing helicopters in the aft portion. Incoming people were placed in squares with corridors separating each square, all their bags were secured on the side of the hanger deck for inspection. The ship began to fill up to where space was becoming an issue, we still had to get our Marines back on board, and the remaining press, and the last embassy people, where are they going to go? We had to separate the Marines from the other refugees.

When the day concluded, 24 helicopters went over the side into the South China Sea, we had over 2,500 new people on the ship, we had our Marines, we had all the Jolly Green Giants on the flight deck, and, we had the news people, embassy people, and we had Air America helicopters (Hueys, Cobras, and all their gear they needed to get out of the country). All aircraft on the flight deck, all people in the hanger deck, refugees forward, some Marines to the rear, all within a rough 24-hour period, though I'm not sure exactly of the time frame.

The embassy was closed; the incoming flights had dropped off, only to be replaced by boats arriving from shore. Unfortunately, we were missing four crewmates we had started out with, the last casualties of the Vietnam war, two young Marines killed defending Tan Son Nhut airport several days earlier and two Hancock rescue helicopter pilots who crashed in the South China sea after many hours flying to save people in the water.

Despite the chaos and uncertainty of the day, it was uncanny how flexible the crew was responding to changing circumstances and how quickly everyone pulled together. I was just amazed. It was Americans coming to the rescue, and there was not, I know as an officer, much bad behavior.

Predominantly there were a lot of people in a situation where they had to take it on themselves personally to do the right thing. I think that was the large unwritten story, and in a way, I couldn't have been prouder to be an American and what we did that day together.

The next day, all the action was over, where are we going to take all these people? Hancock was directed to go back to Subic Bay. A four or so day cruise across the South China Sea. Five thousand people onboard that needed to be fed, a place to sleep, medical attention, sanitary needs. Men, women, and children onboard in the days when only men served on Navy ships.

The people were a cross section of society, all ages, they had nothing. I remember thinking "This is not the elite." I don't care what the politicians say, that, "Oh, the elite that was associated with us, they're the only ones we rescued, and we left everyone else."

It was a cross section of humanity. I remember this elderly woman and this elderly man, they must have been five foot tall and maybe 89 pounds apiece, the black pajamas, the peasant hat, they had no idea where they were, we could have been a spaceship taking them to Mars, they were just bewildered. There were the elderly, there were young, babies, lots of families. Military people in their uniforms, obviously despondent. Families. Moms taking care of their kids, in eerily quiet shock, no yelling, no screaming.

Eating and sleeping on Hancock was challenging, with so many people and absolute zero privacy, the heads were packed 24/7, everyone used everything, the ship began to smell like a Southeast

Asian City. No one complained. We fed them. The hangar deck was organized, they'd get up in a group from their assigned square, and they'd move over, we cleaned it and moved them back. Everyone cooperated.

I have two very distinct memories from the return trip:

One was, there was a woman news reporter who was evacuated recording the news, the next day onboard doing her job. She was walking around on the hangar deck, she was tall, Nordic looking in a white pantsuit, blondish hair, in the middle of all these forlorn people sitting in their assigned squares, looking down, talking into this recorder, dispassionately recording her observations. A striking contrasting image. I said to myself, "How the hell could you do that? How can you do this job? The dispassion. The contrast between her and the refugees was staggering."

I said to myself, "I don't like you and everything you represent. How can you do this?" Not a fair indictment on my part, I'm sure.

Secondly, I volunteered to help return people's belongings to them enroute back to Subic Bay before we docked. The majority had next to nothing except personal items. But here was one guy I remember, he was well fed, blue sports jacket, late 30's, had a yellow hard shell Samsonite briefcase. We had to inspect everything for contraband before we returned their personal items. I opened the briefcase, and it's loaded with stacks of freshly minted Vietnamese money!

He spoke English, I looked at it, and he looked at me, and I said, "You know," he interrupts and says, "I know." I respond haltingly, "This isn't worth anything." And he said, "No." I said, "Can I have a few bills?" He gave me a couple of bills, which I still have to this day. Very emotional, I have zero idea of his background and how he got all that cash, but he left his country with all that money, thinking he was secure, and he realized at that moment, he had nothing but the sport jacket on his back!

We returned to Subic and that was the conclusion of Operation Frequent Wind. There was no news. I didn't know the big picture, and in fact, I didn't really know what went on until I came back and read a Time Magazine that my mom had saved for me.

Since we were at the beginning of our cruise after all the non-ships company had departed, Hancock went back to its regular scheduled duty for the remainder of our cruise, uneventful and routine.

We didn't return to our homeport in Alameda until mid-October.... When the ship was secure to the pier, the Chief, per standard procedure, pulled down the flag from the main mast. I realized that the flag would never fly again due to the impending decommissioning. I asked the chief to give it to me since it had flown continually during the evacuation. I have had the flag ever since.

EPILOGUE

When we returned in mid-October 1975, five months had passed.

In the mind of the country, Vietnam was long gone and long for-gotten. Nobody wanted to talk about it. No one knew anything about it, it was ancient history, and no one was interested except maybe, "Yeah, yeah the helicopter thing over the side, yeah cool." Onto something else. I did not want to dwell on it myself....

Thoughts of Vietnam have intruded over the years. I often tell my peers I am just like them, except for seven days. Years later, my wife and I were watching the film *The Deerhunter* and right at the end of the film is actual footage from Hancock and it's the woman reporter I remember from the Hanger Deck delivering a line about Vietnam which apparently made her career. I still resent her: https://clip.cafe/the-deer-hunter-1978/hilary-brown/

On the 25th anniversary of the evacuation, I was sitting on my current back patio, and I got a call from a shipmate I was close with, "Bob, it's 25 years...", we catch up. He is out in Idaho doing what he wanted to do post USN, we wished each other well, and have not spoken since....

I joined the VFW but have not attended any meetings. In 2015, I noticed that Rory Kennedy had produced a documen-tary called *Last Days in Vietnam*, https://www.youtube.com/watch?v=RTWX-BB4aAA .

Upon watching, I was delighted to find the film exactly tracks my experience. The Naval part of the story focuses on a destroyer USS Knox but was very reminiscent of what we experienced on USS Hancock, but on a much larger scale....

Retired now, as I get older, I start to wonder what to do with my Flag. I noticed in the *Dunwoody Crier* activity related to Viet-nam and the monument in Brook Run Park where I walk with my wife. A young couple of Vietnamese descent moved into the house across the street from me, with two young children. They were both born in the US, are highly educated, living the American dream, their parents were refugees. They are a close-knit family and great neighbors. We don't speak about the war.

I decided to attend the 50th anniversary celebration in Brook Run Park with my flag, but somehow missed the connection, but read the follow-up article about the event in the Dunwoody Crier. I reached out to the people in Dunwoody who connected me to John Butler. We had lunch where he convinced me to come to a meeting which I did attend... my first ever veteran event I have attended since the day I got out of the Navy. I am now a proud member of Atlanta Vietnam Veterans Business Association.

I think after attending, I now know why I have the Flag and what it represents.

POST-VIETNAM EXPERIENCES

ROBERT O. "BOB" BABCOCK

THE REST OF MY NURSE STORY

If you read the story I wrote about *Lieutenant Dexter, RN*, here is "the rest of the story."

As with most Vietnam veterans, I came home, got out of the Army (the day after I got off the Freedom Bird in Oakland), and got on with my life. After working in Chicago for a year in a job I hated, I moved to Kansas City and started a 34-year career with IBM in September 1968.

I kept my eyes on the news from Vietnam, especially the limited news about the 4th Infantry Division fighting on the Cambodian border. News was sparse, it seems reporters liked to stay close to Saigon and other cities close to the coast… not out in the jungle boonies where 4ID was.

I seldom talked about Vietnam (nobody cared, or so it seemed), and my job as an IBM rookie salesman kept me busy, along with starting our Family.

When I did get refocused on Vietnam after my trip to the Wall when the Three Men Statue was dedicated on Veterans Day 1984, and the 20th anniversary of a key event in Vietnam (for me) on November 20, 1986, I started looking for people I had served with.

I found a lot of them who are great friends today. Also, Lieutenant Dexter was someone I wanted to find and thank her for what she did for wounded troops. I had no clue how to find her, knowing she had likely gotten married and changed her name.

I traveled to the Wall virtually every year during the late 1980s and 1990s. (It may seem strange that IBM always had a reason for me to be in or around Washington, D.C. on all Veterans Days back then—smile). In 1992, a big focus was on the upcoming dedication of the Vietnam Women's Memorial on Veterans Day 1993.

At the hotel where I stayed was a table with women Vietnam veterans advertising their memorial. I stopped there and asked if any of them knew a nurse who had served in Vietnam with the name of Dexter. A helpful nurse/veteran said she didn't personally know her, but if I would give her my mailing address and why I wanted to meet her, she would check their membership list and send her a letter (before email) giving her my mailing address. If she wanted to contact me, it would be her choice.

The next month, in December, I got a letter from Judy Dexter Richtsmeier. She told me how much she appreciated my reaching out to her and thanking her for her service at 18th Surgical Hospital in Pleiku, Vietnam. I immediately responded, we talked on the phone, and she told me that she and her husband Ron were 'thinking' about going to the Wall for the dedication of the Women's Monument.

I put a full court press on, encouraging her to go to the Wall. Finally, she agreed she and Ron, her Vietnam Huey pilot husband (they met in the Pleiku hospital when he was wounded), and her Vietnam nurse sister would be there.

Without all the details, suffice it to say that we had a fabulous time at the Vietnam Women's Memorial dedication. She and her sister insisted that Ron and I walk in the parade with them. As it turned out, we were only about a dozen people behind the leaders of the parade.

We went our separate ways, traded Christmas cards for a few years, got on with our lives, and never talked again. Today, as I was typing this, I did a Google search and found that Nurse Judy Dexter Richtsmeier died at age 76, on January 21, 2020, reaching the end of a courageous 40-year battle with lupus. She retired as a full colonel Army nurse in the Army reserves and is buried in

Arlington National Cemetery. You can bet I will visit her grave the next time I go to D.C.

ATLANTA VIETNAM VETERANS BUSINESS ASSOCIATION, WASHINGTON, DC TRIP MARCH 27-29, 2025

FOREWORD: In 2018, a total of 50 members/spouses from the AVVBA traveled to Washington, DC to be part of the first Vietnam War Commemoration Event to be held at the Vietnam Veterans Memorial Wall. At that time, we all decided that we would return for the final Vietnam War Commemoration event in 2025. This story commemorates the events and activities of that trip.

On March 29, 2025, the Vietnam War Commemoration office held an event to honor all Vietnam veterans and their families for their dedicated honorable service and sacrifice to coincide with the 50th Anniversary of the end of the Vietnam War. On that Saturday, AVVBA was invited to participate in the Commemoration office's wreath-laying event at the Wall, honoring the legacy of all Vietnam veterans.

On March 27, 27 AVVBA members and 23 spouses/significant began arriving in Washington to participate in events/activities associated with this special Vietnam War Veterans Day. Everyone stayed at the Embassy Suites by Hilton Alexandria, Old Town.

On Friday, March 28, we headed to the Pentagon for the purpose to visit the Vietnam War Commemoration exhibit. Commemorating the 50th anniversary of the Vietnam War, the Department of Defense and the United States of America Vietnam War commemoration have created a permanent exhibit to thank and honor the service and sacrifice of Vietnam veterans and their families. This award-winning exhibit tells the story of the United

States involvement in the Vietnam War through the timeline of events, artifacts, history photographs and video footage.

The Staff of the DOD commemoration office were our tour guides and provided insight to the exhibits as well as taking photographs since we were not authorized to take photographs in the Pentagon. CDR Brian Wierzbicki provided each of our members with a commemoration challenge coin.

After completing our tour of the Vietnam War exhibit, we were escorted on a tour through the Pentagon particularly in the area that was impacted by the 9/11 attack.

Just outside the Pentagon where the jetliner struck, a $22 million memorial sits on two acres of land. Dozens of crepe myrtle trees surround the serene landscape. There are 184 memorial benches dedicated to each of the victims, and they're organized in a timeline of their ages, from the youngest victim, 3-year-old Dana Falkenberg, to the oldest, 71-year-old John Yamnicky.

After lunch, we went to Arlington National Cemetery. Approximately 27 to 30 funerals are conducted each weekday plus another six to eight are conducted on Saturday for a total of between 141 and 158 services per week. The property is the resting place for more than 400,000 active-duty service members, veterans, and their families. It is a place to remember the men and women who serve their nation; all who visit gain a sense of their sacrifice. Thousands of brave men and women have been laid to rest in Arlington, among them are some of America's most honored heroes.

At approximately 2:45 PM we assembled at the Tomb of The Unknown Soldier for the Changing of the Guard and Wreath Laying ceremony. In 1932, The Tomb of the Unknown Soldier was completed and is the burial site of soldiers whose remains were unidentifiable from World War I, World War II, Korean War, and one Vietnam soldier.

Tomb Guards at the Tomb of the Unknown Soldier are volunteer, enlisted, United States Army soldiers (men or women) assigned to the "3d U.S. Infantry Regiment" also known as "The Old Guard" (TOG). TOG soldiers who have the Military Occupational Specialty (MOS) of 11B basic infantry or 31B military police are eligible to volunteer and apply to serve as Tomb Guards.

Tomb Guards make it their goal to earn the Tomb Guard Identification Badge (TGIB). The Tomb Guard Identification Badge (TGIB) is awarded after the Tomb Guard Sentinel passes a series of tests, including one on the history of Arlington National Cemetery. Tomb guards are always equipped with a rifle and bayonet as well as a sidearm. Guards are currently equipped with the M14 rifle and the Sig Sauer P320 M17 9mm.

There is a meticulous routine in Walking the Mat that the guard follows when watching over the Tomb and crypts:

- The tomb guard marches 21 steps south down the 63-foot-long (19 m) black mat laid across the Tomb.
- Turns and faces east, toward the Tomb, for 21 seconds.
- Turns and faces north, changes weapon to the outside shoulder and waits 21 seconds.
- Marches 21 steps up the mat.
- Turns and faces east for 21 seconds.
- Turns and faces south, changes weapon to the outside shoulder and waits 21 seconds.
- Repeats the routine until the soldier is relieved of duty at the Changing of the Guard.

Note, twenty-one (21) was chosen because it symbolizes the highest military honor that can be bestowed: the 21-gun salute. A 21-gun salute at a funeral is a military honor bestowed upon individuals of high rank or those who have made significant contributions to the nation, typically including the President, ex-Presidents, and some military officials.

After each turn, the guard executes a sharp *Shoulder Arms* movement to place the weapon on the shoulder closest to the visitors to signify that the guard stands between the Tomb and any possible threat.

Out of respect for the interred, the sentinels command silence at the tombs. If the guard walking the mat must vocally confront a disturbance from spectators, or a threat, the routine is interrupted and remains so until the disturbance is under control. The sentinel will exit the mat, place the weapon in port arms position, and confront the disturbance. Once under control, the sentinel then walks on the pavement to the other side of the mat, turns to shoulder arms, and resumes the routine from the point of interruption.

Changing of the Guard occurs while Arlington National Cemetery is open, during the day in the summer months from April 1 to September 30, the guard is changed every half hour. During the winter months, from October 1 to March 31, the guard is changed every hour. After the cemetery closes to the public (7 p.m. to 8 a.m. April through September, and 5 p.m. to 8 a.m. October through March), the guard is changed every two hours.

Wreath Laying at the tomb of the Unknown soldier ceremonies are limited to two per groups per day, with a maximum of four participants in the ceremony. AVVBA was honored to be selected to lay our wreath at the Tomb of the Unknown Soldier. AVVBA members lined up in our typical *At Ease* formation in a special viewing area.

The wreath laying ceremony follows a strict protocol as established by the Honor Guard. The process involves a briefing and confirming with the wreath laying participants their role and the procedure. Following the briefing by the Honor Guard Sergeant, the participants and Honor Guard Sergeant marched down the steps to an area directly in front of the tomb where the wreath is standing.

An Honor Guard member held the wreath and moved backwards toward the tomb of the unknown soldier where two members, Skip Bell and Art Katz placed their hands on the wreath. Max Torrence and Kurt Mueller remained in their position until such time the wreath had been placed in front of the tomb at the designated location.

At this point the Honor Guard Sergeant commanded *Present Arms,* at which time taps was played. Afterwards *Order Arms* was given and the participants Skip Bell, Art Katz, Max Torrence and Kurt Mueller in concert with the Honor Guard Sergeant did an *About Face* and marched back up the steps, at which time the Honor Guard Sergeant acknowledged what a great job we did, and thanked us for our service.

In the middle of the wreath lane ceremony, during the playing of taps, four Air Force T-38s flew over representing the missing man formation. The honor guard members were surprised that we were able to get the timing of the jets right during the middle of our ceremony.

At the end of the wreath laying event, we met with several high school students who were very interested in talking with us and wanted to take pictures with our group.

Day 3, Saturday, March 29—after breakfast we all boarded the motor coach headed to the National Mall and The Vietnam Veterans Memorial (more commonly referred to as 'The Wall'.) Prior to the event we gathered our group in front of the Lincoln Memorial in preparation for the presentations from the commemoration office and dignitaries. At 10:00 am we attended the 8th Anniversary of the National Vietnam Veterans War Day event, which was hosted by the Vietnam War Commemoration office.

Mark Franklin was the master of ceremonies. In attendance

was Major Phil Pershing, Regimental Chaplain, Major General Ed Chrystal, US Army, Director of the Vietnam War Commemorative Office and the Honorable Doug Collins, Secretary of Department of Veteran Affairs. In Major General Ed Chrystal's speech, he specifically acknowledged our group was present at the event. Doug Collins was the keynote speaker.

After the speeches, there was a military procession and wreath laying event at the Wall conducted by the dignitaries from the previous event, a second wreath with individuals to honor our national warriors, and a third wreath to honor the warriors of the families.

Following *Amazing Grace* played by the bagpiper, our group marched in formation to the Wall, assembled at the Wall, placed our wreath, and then rendered our hand salute. Following this event, professional photographs were taken with Doug Collins, Secretary of Department of Veteran Affairs and Major General Ed Chrystal.

Our group then stayed in the area and met other groups and organizations. All in all, it was a great and memorable trip, enjoyed by all who attended.

THE VIETNAM WAR MEMORIAL OF GEORGIA

When Vietnamese American Community of GA (VAC-GA) President TraMy Nguyen and Atlanta Vietnam Veterans Business Association (www.AVVBA.org) President John Butler first met in 2020, TraMy shared her dream of creating a memorial to honor the lives of both the Vietnamese and U.S. Soldiers who sacrificed their lives fighting for freedom during the Vietnam War.

TraMy's specific desire was to create a memorial to honor all 400,000 soldiers who sacrificed their lives fighting in the Vietnam War, especially those from South Vietnam and the U.S., but also including the other six countries who lost soldiers while serving in Vietnam.

Tens of thousands of the 160,000 Vietnamese Americans living in Georgia were born in South Vietnam and came to America after the fall of Saigon on April 30, 1975. Many risked their lives to escape to America. Their families have endured indescribable suffering at the hands of the brutal Communist authorities from North Vietnam who invaded their sovereign nation.

TraMy was born in Vietnam, and her father served in the South Vietnamese Army during the war. After the Fall of Saigon, he was arrested and spent several years in a Communist "reeducation camp" before finally being released, then escaping with his family to America.

Nearly all the VAC-GA members are either Vietnamese Refugees themselves or family members of Vietnamese Refugees. Even though membership in the organization is not carefully recorded or managed, they often have several thousand gather to participate in their events.

All the 250 members of the AVVBA served in the U.S. military in Vietnam during the war. In one way or another, we all joined our South Vietnamese allies in defending their country from the illegal invasion of the Communist country of North Vietnam. The desire runs deep in the Vietnamese American people and in the American soldiers who fought in Vietnam to honor all of the soldiers who sacrificed their lives fighting for freedom from Communism in Vietnam.

At first, TraMy's dream seemed impossible. However, as TraMy and John persisted in pursuing options and ideas with each other, the dream evolved into a potential reality. In April of 2022, a project team was formed who began working in earnest to transform the dream into the memorial that exists today. From the beginning, the members of the project team included representation from both the Vietnamese American Community and the Atlanta Vietnam Veterans Business Association.

Over the next few months, multiple sites were explored, each of which ended up not being feasible for one reason or another. Finally, in the Fall of 2022, one of the project team members discovered Brook Run Park in Dunwoody, GA. It was perfect!

At AVVBA's board meeting in September of 2022, the leadership team was asked if anyone knew a member of Dunwoody's City Council. Board member Jay Pryor said he and his wife, Barbara, knew Dunwoody Mayor Lynn Deutsch and her husband. Jay offered to set up a meeting with Mayor Duetsch, which resulted in Jay instantly becoming a member of the project team!

In October of 2022, two members of the project team met with Dunwoody Mayor Lynn Deutsch. She was very receptive to the concept and recommended a process to be followed to obtain the support of the City Council for the project.

Following numerous meetings with City Council members, Parks and Recreation Staff, and the City Manager, the project team presented a proposal to the City Council in May of 2023. In June of 2023, the City Council voted unanimously to accept the project team's proposal to build a privately funded memorial in Dunwoody's beautiful Brook Run Park.

Design of the memorial components, the overall memorial itself, and the layout of the memorial site in relation to the overall site within the park was done in concert with Dunwoody's Parks and Recreation Director, initially Brent Walker, then Rachel Waldron after Brent was hired by Sandy Springs.

The design included two 12' wide x 8' tall memorial walls to honor the soldiers who died in the war. The front of the left wall would feature information about U.S. units and soldiers. The words would be laser etched in English, then in Vietnamese. The front of right wall would feature information about Vietnamese units and soldiers. Those words would be laser etched in Vietnamese, then in English.

The backs of the two memorial walls would have information about the war itself as well as information about the soldiers from the other six allied countries; all etched in both Vietnamese and English. An American flag would fly above the left memorial wall and the flag of South Vietnam would fly above the right memorial wall.

The entrance walkway would have two donor walls on each side. Each of the two donor walls on each side would be 16' long and 4' tall and positioned at a 45-degree angle so the names of each donor sandblasted on those walls could be easily read from the sidewalk.

A life-sized statue of two soldiers would be mounted on top of a 4' tall black granite base positioned in the middle of a 30' x 40' oval "plaza" with the memorial walls toward the back and granite benches placed around the front on either side of the entrance sidewalk. The statue would consist of a U.S. infantry soldier standing next to a South Vietnamese infantry soldier, both on alert as if they are on patrol together.

In order for the soldiers to appear to be life-sized when standing 4' above the ground, each soldier would be about 12" taller than their normal average height. The U.S. soldier would then need to be 7' tall and the Vietnamese soldier would be 6' 6" tall. The details of their uniforms and equipment would be consistent with what was most commonly worn and used in 1968 and 1969, the years of the highest casualties and largest allied involvement of the war. Both Vietnamese and U.S. veterans who served in infantry units in Vietnam during those years would be consulted to ensure accuracy of uniform and equipment details.

Muted lighting would project onto the statue, the flags, the donor walls, and onto both sides of the memorial walls during the night.

Names, dates and other information about veterans who served in Vietnam would be sandblasted onto the backs of each bench. The sandblasting of that information would be funded by family members in honor of or in memory of their loved one.

A 3-D conceptual rendering of the memorial and a virtual flyover video were both professionally produced in August, 2023. Both significantly enhanced the project team's ability to describe the scope and significance of what would be created, and to increase interest in funding.

On September 12, 2023, a groundbreaking ceremony was conducted by the project team and was attended by Dunwoody's Mayor and City Council, several dignitaries, and dozens of members of both the VAC-GA and the AVVBA.

Within a few months, all of the $1.25 Million project cost was raised from over 400 individuals and small businesses plus 10 local non-profit organizations who donated $500,000 in cash. Nearly all of these donations were made by members and friends of the Vietnamese American Community and the Atlanta Vietnam Veterans Business Association.

The other $750,000 provided was "in-kind" contributions contributed from several local businesses, most of which were veteran-owned including three owned by AVVBA members. Two of them were responsible for over $500,000 of those contributions.

One of those two businesses was Memories Last Forever Monument company, owned by AVVBA member Sammy Robinson who served as an Army Infantry soldier in Vietnam. His significant connections in the monument industry, and experience in designing and building other veterans memorials enabled him to contribute not only invaluable guidance throughout the project, but also provide the memorial walls, the donor walls, the statue and base, and all of the laser etching and sandblasting at huge savings to the project.

Another of those two businesses was Willow Construction, owned by AVVBA member Alan Gravel and his son Mark. Alan served as an Air Force pilot in Vietnam. Alan, Mark, and their amazing team of highly skilled workers provided all the considerable site work required for the project.

Since the site was in the middle of a busy, upscale park in one of the most prestigious communities in the Atlanta area, their heavy, rough work had to be accomplished in a delicate environment within the constant scrutiny of "interested" neighbors, many of whom had lots of questions and suggestions!

Most of their machines were transported to the side on huge flat-bed semi-trailer trucks. The largest machine involved was an enormous 20-ton crane that had to pick up a 13,000 pound block of granite from the bed of a flatbed truck, carry it up and over mature bushes to the middle the site, and set it down in precisely the correct position in the center of the memorial plaza to become the base for the statue of the two soldiers.

The heavy straps under the massive granite block had to be removed once the block was in place. No one watching this fascinating process could imagine how those straps were to be removed after the block was sitting flush onto the flat concrete slab. Not to worry, Sammy and his crew went to a local convenience store and bought four 20-pound bags of ice.

They positioned the bags of ice on the slab within the margins of where the granite block would be sitting. The block was lowered onto the ice bags. The straps were removed. The ice melted. The granite block ended up on the plaza with the now empty (now extremely thin) bags forever beneath it!

The Memorial was dedicated and opened to the public on October 5, 2024, with about 400 in attendance. Live music, including Taps, was provided by University of North Georgia's Golden Eagle Cadet Band. A 21-gun salute was provided by the Marine Corps League. Military vehicles including a Huey helicopter that had served in Vietnam were on display. A hot air balloon displaying huge South Vietnam and American flags flew nearby.

Two retired Army Generals provided keynote speeches: one by U.S. Army Lt. General (Ret.) Ron Helmly who was born in the U.S. and served two tours in Vietnam; and the other by U.S. Army Maj. General (Ret.) Lapthe Chau Flora who was born in Vietnam and escaped to the U.S. as a child. General Flora is the only Vietnamese "boat person" who became a U.S. Army General.

As far as we know, this is the only memorial in existence jointly produced and funded by a partnership between Vietnamese Refugees and U.S. Vietnam Veterans that honors both Vietnamese and U.S. Soldiers who were killed during the Vietnam War with a life-sized statue of a South Vietnam Soldier standing next to a U.S. Soldier as the centerpiece of the memorial.

The Vietnam War Memorial of Georgia is a 501(c)(3) non-profit corporation responsible for the on-going well-being of the memorial, interaction with future donors, and memorial enhancements, all in coordination with the City of Dunwoody. The board consists of a balanced representation of Vietnam Veterans and Vietnamese Americans with TraMy Nguyen and John Butler serving as co-chairs.

As of this printing, spaces for donors to be engraved on donor walls and benches in honor or memory of Vietnam Veterans are still available to be purchased at www.VNWarMemorial.org.

The Vietnam War Memorial of Georgia is pictured on the back cover of this book.

SOUTH VIETNAM

www.ingramcontent.com/pod-product-compliance
Lightning Source LLC
Chambersburg PA
CBHW061236220326
41599CB00028B/5441